Liquid Crystal Polymers I

Editor: M. Gordon
Guest Editor: N. A. Platé

With Contributions by
P. J. Flory, J.-I. Jin, R. W. Lenz,
Ch. K. Ober, S. P. Papkov,
I. Uematsu, Y. Uematsu

With 79 Figures and 11 Tables

Springer-Verlag
Berlin Heidelberg GmbH
1984

ISBN 978-3-662-16008-4   ISBN 978-3-540-38748-0 (eBook)
DOI 10.1007/978-3-540-38748-0

Library of Congress Catalog Card Number 61-642

© Springer-Verlag Berlin Heidelberg 1984
Originally published by Springer-Verlag Berlin Heidelberg New York Tokyo in 1984
Softcover reprint of the hardcover 1st edition 1984

2152/3020–543210

Editors

Editorial

With the publication of Vol. 51, the editors and the publisher would like to take this opportunity to thank authors and readers for their collaboration and their efforts to meet the scientific requirements of this series. We appreciate our authors concern for the progress of Polymer Science and we also welcome the advice and critical comments of our readers.

With the publication of Vol. 51 we should also like to refer to editorial policy: *this series publishes invited, critical review articles of new developments in all areas of Polymer Science in English (authors may naturally also include works of their own)*. The responsible editor, that means the editor who has invited the article, discusses the scope of the review with the author on the basis of a tentative outline which the author is asked to provide. Author and editor are responsible for the scientific quality of the contribution; the editor's name appears at the end of it.

Manuscripts must be submitted, in content, language and form satisfactory, to Springer-Verlag. Figures and formulas should be reproducible. To meet readers' wishes, the publisher adds to each volume a "volume index" which approximately characterizes the content.

Editors and publisher make all efforts to publish the manuscripts as rapidly as possible, i.e., at the maximum, six months after the submission of an accepted paper. This means that contributions from diverse areas of Polymer Science must occasionally be united in one volume. In such cases a "volume index" cannot meet all expectations, but will nevertheless provide more information than a mere volume number.

From Vol. 51 on, each volume contains a subject index.

Editors   Publisher

Table of Contents

Molecular Theory of Liquid Crystals
P. J. Flory 1

Polypeptide Liquid Crystals
I. Uematsu, Y. Uematsu 37

**Liquid Crystalline Order in Solutions of
Rigid-Chain Polymers**
S. P. Papkov 75

**Liquid Crystal Polymers with Flexible Spacers
in the Main Chain**
Ch. K. Ober, J.-I. Jin, R. W. Lenz 103

Author Index Volumes 1–59 147

Subject Index 155

Molecular Theory of Liquid Crystals

Paul J. Flory

Department of Chemistry, Stanford University, Stanford, CA 94305, USA

The lattice theory of liquids consisting of rodlike molecules is presented and discussed with emphasis on polymers exhibiting nematic or cholesteric liquid crystallinity. Steric repulsions between the solute particles are principally responsible for order in lyotropic liquid-crystalline systems. In the case of rigid rods, the axial ratio of the particles governs the concentration at which separation of a nematic or cholesteric phase sets in. For semi-rigid chains such as those of cellulose and its derivatives, the axial ratio of the Kuhn segment is the relevant parameter. These and other predictions of the lattice theory are confirmed by numerous experiments. Liquid crystallinity may be promoted by orientation-dependent intermolecular attractions between extended chain molecules. Such forces originate in the anisotropy of the polarizabilities of groups, e.g., phenylene, in the main chain. They may be especially important in thermotropic melts and concentrated solutions.

1 Introduction 2

2 Binary Lyotropic Systems Consisting of Hard Rodlike Particles and a Diluent   4
   2.1 Formulation of the Partition Function 4
   2.2 Order vs. Disorder. 7
   2.3 Biphasic Equilibrium. 9

3 Binary Lyotropic Systems. Comparison of Theory with Experiment 11

4 Multicomponent Systems 14
   4.1 Polydisperse Rods 14
   4.2 Mixtures of Rods and Random Coils 16

5 Semirigid Chains 18
   5.1 Cellulose and its Derivatives 20
   5.2 Other Polysaccharides 21
   5.3 Polymers Comprising Both Rigid and Flexible Units or Sequences . . . 22
   5.4 Conformational Changes Coupled with the Isotropic-Nematic Transition   24

6 "Soft" Intermolecular Interactions 25
   6.1 Isotropic Interactions 25
   6.2 Orientation-Dependent Interactions 27

Appendix:
   A Lattice Theory for Hard Rods with Exact Treatment of the Orientation
     Distribution 31
   B Rods with Orientation-Dependent Interactions 33
   C The 1956 Approximate Treatment. 34

References 35

Advances in Polymer Science 59

1 Introduction

Asymmetry of molecular shape is a feature common to all substances that exhibit liquid crystallinity [1-4]. Low molecular nematic and cholesteric liquids usually consist of rodlike molecules having axial ratios in excess of three. Platelike molecules, or particles, also may adopt states of mesomorphic order that exhibit the characteristics associated with liquid crystallinity. The most definitive of these characteristics as manifested in typical nematic liquid crystals is uniaxial order which, though imperfect, is of long range; the preferred axis is maintained throughout the domain of the liquid crystal whose dimensions are macroscopic. Liquid crystals differ from true crystals in the absence of long-range molecular order in directions transverse to the domain axis and in the imperfection of molecular alignment along this axis. In other respects, nematic liquid crystals bear a close resemblance to ordinary liquids. They exhibit diffuse X-ray scattering patterns, their equations of state are nearly identical with those of the same substance in the isotropic state, and their fluidities at low rates of shear are qualitatively similar to those of ordinary liquids.

Polymers that exhibit liquid crystallinity, either in the melt or in their solutions, typically consist of comparatively rigid structures that confer high extension on the backbone of the macromolecule. This molecular feature is obviously conducive to the axial order that is the mark of a nematic fluid[1].

Intermolecular attractive forces also may contribute to the stabilization of the liquid crystalline state if these forces depend on the relative orientation of the interacting molecules, or, more precisely, of those portions (segments) of them that are in close proximity. Anisotropy of the polarizability of the molecule, or of its constituent groups, is the molecular feature that is required to render the intermolecular forces orientation-dependent. Just as the scalar, or mean, polarizability determines the strength of the London dispersion forces between isotropic molecules, anisotropies of the polarizability tensors of the interacting molecules confer a corresponding anisotropy on the dispersive attractions between "optically anisotropic" molecules. Parallel molecular alignment may be energetically favored on this account. Forces of this nature comprise the basis for the Maier-Saupe [5] theory of the nematic liquid crystalline state. They appear to play an important role in nematogenic molecules, both low molecular and polymeric, that contain aromatic groups structurally aligned with the molecular axis. The p-phenylene group, for instance, exhibits a substantially larger polarizability along its main axis compared to the transverse directions. The presence of this or other arylene groups in virtually all low molecular nematogens gives special prominence to orientation-dependent London forces in this broad class of liquid crystalline substances. Orientation-dependent interactions in polymers containing arylene groups as integral members of the chain backbone may likewise promote liquid crystallinity. The dominant molecular characteristic responsible for liquid crystallinity appears invariably, however, to be asymmetry of molecular shape.

---

1 A special class of polymeric liquid crystals is obtained by attaching rigid side chains to a flexible backbone. Inasmuch as it is the side chains that engage in formation of the liquid crystalline domains, these systems are more closely akin to low molecular liquid crystals. Polymeric liquid crystals of this type are discussed by Rehage and Finkelmann in Vol. 60 of this series.

Structures of liquids in general are dominated by influences of intermolecular repulsions. Intermolecular attractions have a comparatively minor effect on the radial distribution function and, in the case of asymmetric molecules, on intermolecular correlations as well. At the high density and close packing prevailing in the liquid state, the spatial arrangement of the molecules of a liquid can be satisfactorily described, therefore, by representing the molecules as "hard" bodies of appropriate size and shape whose only interactions are the excessive repulsions that would be incurred if one of them should overlap another. Once the liquid structure has been characterized satisfactorily on this basis, one may take account of the intermolecular attractions by averaging them over the molecular distribution thus determined. Mean-field theories are useful in this connection.

Lattice methods are well suited to treatment of the intermolecular, or "steric", part of the configuration partition function for a system comprising particles or molecules in which volume exclusion plays its usual, dominant role. In principle, these well established methods are no less applicable to systems consisting of species of asymmetric shape, although certain modifications of the conventional procedures are required in order to accommodate highly asymmetric molecules on a lattice when they maintain a degree of orientational disorder. Lattice theory has been adapted to this purpose and the practicability of this approach has been demonstrated [6, 7].

Lattice methods suffer the disadvantage of apparent lack of rigor that is an inevitable consequence of adoption of an artificial model, in this instance a regular array of sites, for the representation of a disordered liquid. Owing to the marked periodicity of the radial distribution function of the liquid over short ranges, unrealities of the model in this regard are not as serious as may at first appear.

The virial expansion has enjoyed greater appeal, especially as applied to lyotropic systems. Onsager's classic theory [8] rests on analysis of the second virial coefficient for very long rodlike particles. It was the first to show that a solution of hard, asymmetric particles such as long rods should separate into two phases above a threshold concentration that depends on the axial ratio of the particles. One of these phases should be anisotropic (nematic), the other completely isotropic. The former is predicted to be somewhat more concentrated than the latter, but it is the alignment (albeit imperfect) of the solute molecules that is the predominent distinction. The phase separation is a consequence of shape asymmetry alone; intervention of intermolecular attractive forces is not required.

An approach based on the virial expansion suffers from the difficulty of evaluating higher coefficients for highly asymmetric particles and from the non-convergence of the virial series at the concentrations required for formation of a stable nematic phase [6, 7]. Lattice methods therefore take precedence over the virial expansion as a basis for quantitative treatment of the liquid crystalline state.

We shall not attempt to review and compare critically various theories of liquid crystallinity in this chapter. Inasmuch as theory based on a lattice model has proved most successful in the treatment of liquid crystallinity in polymeric systems, we shall present an abbreviated account of that theory confined to its essential aspects. The versatility of this theory has permitted its extension to polydisperse systems, to mixtures of rodlike polymers with random coils and to some of the many kinds of semirigid chains. These ramifications of the theory will be discussed in this chapter and comparisons of theoretical predictions with experiments will be summarized.

Finally, the effects of intermolecular interactions will be considered. These may arise from isotropic interactions that are commonplace in mixtures in general or from orientation-dependent energies cited above. Both types of interactions are easily incorporated in the theory based on the lattice model.

2 Binary Lyotropic Systems Consisting of Hard Rodlike Particles and a Diluent

2.1 Formulation of the Partition Function

Conventional lattice methods can be adapted to treatment of a system of rigid, rodlike particles, or molecules, by the device illustrated in Fig. 1 [6]. The particle shown in Fig. 1a is oriented at an angle ψ with respect to the preferred axis of the surrounding domain. One of the principal axes of the cubic lattice is aligned with this axis. The particle comprises x isodiametric segments, each of a size that will occupy one cell of the lattice; it follows that x so defined is also the axial ratio of the particle. In order to accommodate the particle on the lattice when it is oriented as in Fig. 1a, we imagine it to be subdivided into y sequences of segments as represented in Fig. 1b, with each sequence oriented parallel to the preferred axis. The parameter y serves as a measure of disorientation of the particle with respect to the domain axis; see Fig. 1b.

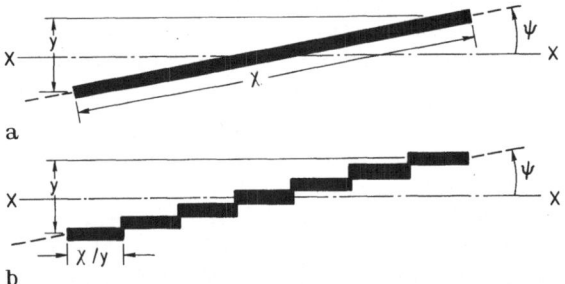

Fig. 1. (a) Rodlike particle oriented at an angle ψ from the preferred axis X $\cdots$ X, and (b) its subdivision into y sequences parallel to this axis as required for accommodation on the cubic lattice [6] (From Flory and Ronca [7])

The analysis depends on evaluation of the number ν_j of situations (i.e., appropriate sequences of empty lattice sites) accessible to molecule j with orientation y_j when $j - 1$ molecules have been added previously to the space, or lattice, in which the mixture is confined. This quantity can be formulated as the product of the number of lattice sites available to the initial segment of the chain and the probabilities that the sites required for each successive segment of the chain are unoccupied and, hence, accessible. The number of eligible locations for the initial segment of the chain is just the total number $n_0 - x(j - 1)$ of vacant sites, where n_0 is the total number of lattice sites. In formulating the probability that the site required by the second segment of the first of the y_j sequences is unoccupied, one must take account of the fact that the preceding site has been established to be vacant. If the second site should turn out to be occupied by a segment of a molecule previously incorporated in the lattice, the occupant must necessarily be the first segment of one of the sequences in that molecule; occupation by a segment beyond the first in a sequence is precluded by the assured vacancy of the preceding site. Hence, the site in question may be one of the $n_0 - x(j - 1)$ vacancies in the lattice or it may be occupied by one of the "first" segments of the $\bar{y}(j - 1)$ sequences present. The required *conditional probability* of vacancy of the site following (in the direction of the preferred axis) one known to be vacant is given therefore by the number fraction N_j of vacancies out of the sum of vacancies and of the $\bar{y}(j - 1)$ sequences from molecules previously introduced; i.e., by

$$N_j = [n_0 - x(j - 1)]/[n_0 - (x - \bar{y})(j - 1)] \tag{1}$$

where $\bar{y}$ is the mean value of y. (Without error or sacrifice of generality, $\bar{y}$ may be maintained constant throughout addition of the total number n_p of rodlike solute molecules.) The same number fraction N_j expresses the conditional probability that the site required for the next segment of the sequence also is vacant; and so on to the end of the sequence.

Accessibility of the site required for the first segment of the following sequence is not contingent on vacancy of a preceding site. Hence, the *a priori probability* P_j of a vacancy is required. It is just the volume fraction of vacancies, i.e.,

$$P_j = [n_0 - x(j - 1)]/n_0 \tag{2}$$

It follows that

$$v_j = [n_0 - x(j - 1)] \, N_j^{x-y_j} P_j^{y_j - 1} \tag{3}$$

$$= [n_0 - x(j - 1)]^x/[n_0 - (x - \bar{y})(j - 1)]^{x-y_j} \, n_0^{y_j - 1} \tag{4}$$

$$= \frac{[n_0 - x(j - 1)]! \, [n_0 - (x - \bar{y})j]!}{(n_0 - xj)! \, [n_0 - (x - \bar{y})(j - 1)]! \, n_0^{y_j - 1}} \tag{4'}$$ [2]

The number fraction N_j exceeds the *a priori* probability P_j of a vacancy to the extent that y is less than x; compare Eqs. (1) and (2). Hence, v_j is maximal for $y_j = 1$ and it decreases with increase in y_j, i.e., with disorientation of rod j.

The "steric" or "combinatory" part Z_{comb} of the partition function for the system follows as the product of the v_j for each of the n_p rodlike molecules incorporated in the lattice; i.e.,

---

2 The bracketed quantities in the numerator and denominator of Eq. (4) raised to the powers x and $x - y_j$, respectively, are approximated by ratios of factorials in Eq. (4'). The approximations are negligible for the large numbers in a macroscopic system. Moreover, of the two equations, Eq. (4) and (4'), the latter should be considered more nearly exact. Hence, neither equation is denoted above as approximate.

$$Z_{comb} = (1/n_p!) \prod_{j=1}^{n_p} v_j \tag{5}$$

the factor $1/n_p!$ having been included in order to eliminate redundant configurations that differ only in permutations of the n_p identical rodlike molecules. It expresses the number of non-overlapping configurations accessible to the system of rodlike particles in the state of disorientation designated by $\bar{y}$. As follows from the remarks above, Z_{comb} increases with orientation; conversely stated, it decreases with $\bar{y}$. The result obtained by substitution of Eq. (4′) in (5) is

$$Z_{comb} = \frac{(n_s + \bar{y}n_p)!}{n_s! n_p! (n_s + xn_p)^{n_p(\bar{y}-1)}} \tag{6}$$

where $n_s = n_0 - xn_p$ is the number of vacant sites that remain. They may be filled by n_s molecules of solvent.

Both the numerator and the denominator of Eq. (6) increase with $\bar{y}$. The denominator increases more rapidly, however. Hence, Z_{comb} decreases with disorientation. This is a direct consequence of the decrease in v_j with y_j already noted. The physical reason for the decrease in Z_{comb} with disorientation is at once apparent; it reflects the increasing expectation of steric overlaps between rods with increasing misalignment.

The formulation of v_j is the fundamental step in the derivation of this essential factor in the partition function [6,7,9]. It rests directly on the validity of the device portrayed in Fig. 1 for rendering the disoriented rod susceptible to representation on the lattice through its replacement by a succession of sequences parallel to the preferred principal axis. That this device should afford an acceptable approximation for the estimation of v_j is intuitively evident; it gains support from consideration of analogous situations.

The complete partition function Z_M is the product of Z_{comb} and the function Z_{orient} that takes account of the orientational distribution; i.e.,

$$Z_M = Z_{comb}Z_{orient} \tag{7}$$

The latter function may be expressed by

$$Z_{orient} = \prod_y (\omega_y n_p/n_{py})^{n_{py}} \tag{8}$$

where n_{py} is the number of rods with disorientation from the preferred axis denoted by y, and ω_y is the fraction of solid angle associated with y. In order to proceed it is necessary to establish the relation between ω_y and y, and to evaluate the distribution of orientations denoted by the ratios n_{py}/n_p for various values of y. Rigorous relations for this purpose were derived by Flory and Ronca in 1979 [7]. They are given in Appendix-A; see Eqs. (A-4)–(A-8). When incorporated in the theory these relations lead to integrals that can only be solved numerically; see Eqs. (A-10) and (A-11).

The required calculations are straightforward [7], but the relationships obtained do not reduce to tractable algebraic expressions that illumine the results of the theory.

A simplification that circumvents these difficulties and facilitates extension of the theory to diverse systems of greater complexity was adopted in the earliest version of the theory offered in 1956 [6]. On the grounds that the form of the distribution of orientations should have little effect on the results, the polar angles ψ of the rods were taken to be uniformly distributed within the range $0 \leqq \psi \leqq \psi^+$, with none exceeding ψ^+. This led to the approximation [6]

$$Z_{orient} \approx (\bar{y}/x)^{2n_p} \qquad (9)^3$$

2.2 Order vs. Disorder

Whereas Z_{comb} decreases with disorientation, Z_{orient} increases. For sufficiently small axial ratios (x) and/or concentrations, the product of these factors, i.e., the partition function Z_M (see Eq. (7)) for the mixture of rods and diluent, is dominated by Z_{orient}. Then Z_M increases monotonically with $\bar{y}$ and the state of complete disorder with $\bar{y} = x$ is the one of maximum stability (see Eq. (C-1) in Appendix-C). As the axial ratio and/or concentration are increased, the decrease in Z_{comb} with $\bar{y}$ becomes more marked. Eventually, Z_M exhibits a maximum followed by a minimum with increase in $\bar{y}$, i.e., with increasing disorder. The maximum marks either a metastable or a stable state with respect to orientation. It is stable only if Z_M at the maximum exceeds Z_M for complete disorder ($\bar{y} = x$). This requires a somewhat greater concentration, or axial ratio, than that at the critical point where the extrema first appear.

These characteristics of the dependence of the partition function on disorder are illustrated in Figure 2 where Z_M is plotted on a logarithmic scale against $\bar{y}$ for rods with axial ratio $x = 100$ at the several volume fractions $v_p = xn_p/(n_s + xn_p)$ indicated with each curve. The curves have been calculated from Z_M taken as the product of Eqs. (6) and (9); see Eq. (C-1) in the Appendix. Through division of Z_M by its value Z_M^0 for complete disorder where $\bar{y} = x$, all curves are brought to the same intercept in this limit. At volume fractions below the critical value $v_p^* = 0.0784$, Z_M increases monotonically with $\bar{y}$. Extrema appear at concentrations beyond v_p^*. The maximum represents the state of equilibrium if $Z_{M, max}$ exceeds Z_M^0, as noted above. Even at the critical point $\bar{y} = \bar{y}^*$ is somewhat less than x/2. It decreases as the concentration is increased beyond v_p^* and hence is substantially less than x/2 for a stable phase (see Fig. 2). Thus, the transition from complete disorder ($\bar{y} = x$) to partial order is predicted to occur *abruptly and discontinuously*. It is important to observe that the degree of disorder, although much lower than for isotropy, remains finite ($\bar{y} > 1$) even at volume fractions (or axial ratios) much beyond the point at which the system becomes anisotropic.

The character of the transition from disorder to (partial) order may be better understood through analysis of relevant characteristics of the partition function. By

---

3 Equation (9) differs from previous renditions [6, 7, 9], according to which $Z_{orient} \approx \bar{y}^{2n_p}$, through choice of the randomly disoriented rods ($\bar{y} = x$) as the state of reference, instead of adopting the state of perfect alignment ($y = 1$) for this purpose. The difference is of no significance in applications of the theory here discussed.

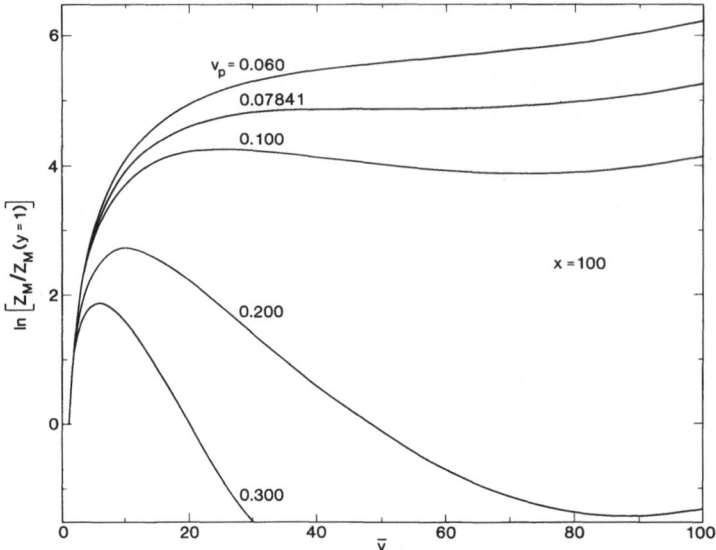

Fig. 2. The partition function Z_M for athermal solutions of hard rods having an axial ratio $x = 100$ shown as a function of the disorder index $\bar{y}$ for the volume fractions indicated. Calculations were carried out according to Eq. (C-1) of the Appendix with $\chi = 0$. The logarithm of the partition function relative to the state of perfect alignment ($\bar{y} = 1$) is plotted on the ordinate

taking the partial derivative of the partition function Z_M with respect to $\bar{y}$ and equating it to zero one obtains the relation

$$\bar{y}/x = 1 - [1 - \exp(-2/\bar{y})]/v_p \tag{10}$$

the 1956 rendition for Z_M (see Eq. (C-1) of Appendix-C) being used for this purpose. Solutions of this equation locate the maximum and the minimum in Z_M with $\bar{y}$ for given values of x and v_p. The former solution denotes the stable (or metastable) state of partial order. Unfortunately, Eq. (10) cannot be solved explicitly for $\bar{y}$ and, hence, used to eliminate $\bar{y}$ from the various relationships that follow from the partition function. It is necessary instead to obtain numerical solutions by solving these relations simultaneously with Eq. (10). The predicted dependence of $\bar{y}$ on v_p and x at orientation equilibrium, described briefly above, may be examined by numerical solution of Eq. (10), however.

A useful semi-empirical approximation for the critical volume fraction v_p^* for incipience of metastable order (see Fig. 2) that follows from analysis [6] of Eq. (1) is

$$v_p^* \approx (8/x)(1 - 2/x) \tag{11}$$

which holds within 2% for $x > 10$. This relation expresses the volume fraction below which solutions of Eq. (10) do not exist. It has been widely misinterpreted as the threshold volume fraction for appearance of a stable anisotropic phase; *cf. seq.*

2.3 Biphasic Equilibrium

The foregoing analysis of anisotropy in a system of rodlike particles ignores the possibility of separation of the system into two phases. Examination of the partition function, or the corresponding reduced free energy — $\ln Z_M$, shows that the critical point discussed above is "bridged" by phase separation, i.e., the critical point occurs in a range of composition (and disorder) that is metastable with respect to a system of two phases, one of them completely disordered and the other partially ordered but with retention of an appreciable degree of disorder. These deductions from the lattice model confirm those obtained earlier by Onsager [8] through analysis of the virial expansion truncated at its second term (see above).

Biphasic coexistence is most conveniently treated by imposing the condition that the chemical potentials μ_s and μ_p (obtained as partial derivatives of the free energy) of the respective components in the two phases must be equal at equilibrium, i.e., that

$$\left.\begin{array}{c} \mu_s = \mu'_s \\ \mu_p = \mu'_p \end{array}\right\} \tag{12}$$

where the prime denotes the anisotropic, or ordered, phase. Expressions for the chemical potentials are given in the Appendix A. Figure 3 shows compositions calculated for the coexisting phases in a binary system consisting of hard rods and a diluent as functions of the axial ratio. The calculations were carried out using the chemical potentials for the ordered, or nematic, phase derived from the exact rendition of the

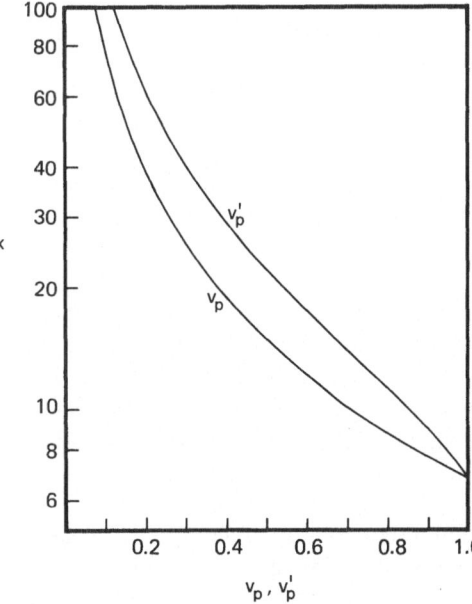

Fig. 3. Compositions of coexisting isotropic and anisotropic phases expressed in volume fractions, v_p and v'_p, respectively, as functions of the axial ratio x of hard rods in athermal solutions. Calculated from the chemical potentials given by the "exact" lattice theory [7]; see Appendix-A (From Flory and Ronca [7])

lattice theory [7]; see Eqs. (A-13) and (A-14). Thermodynamic relations applicable to the isotropic phase are identical with those given by conventional theory of polymer solutions; see Eqs. (A-15) and (A-16). The volume fractions v_p and v_p' in the isotropic and nematic phases coexisting at equilibrium are shown in Fig. 3 by the coordinates on the abscissa of points on the respective curves for the given value of x plotted along the ordinate. The difference in composition of the two phases is relatively small. The biphasic gap narrows with decrease in x. The two curves merge at x = 6.42, the calculated minimum axial ratio for stable nematic order in a neat liquid consisting of hard rods [7]. The phases that coexist at any larger value of x are distinguished more markedly by the long-range order prevalent in one of them than by the difference between their concentrations, the latter difference being minor by comparison.

Characteristics of the biphasic equilibria calculated from the lattice theory are summarized in Table 1. Results obtained for the volume fractions v_p and v_p' at coexistence, for their ratio v_p'/v_p, for xv_p and for $\bar{y}/x$ according to the 1956 approximate version of the theory [6] are given in columns 2, 3 and 4 for the neat fluid, for rods with x = 20 and in the limit x = ∞, respectively. Corresponding calculations from the exact version of the theory [7] are given in the last three columns. The latter calculations yield somewhat lower volume fractions for the coexisting phases. The ratios v_p'/v_p are smaller; in the limit x → ∞ this ratio is 1.465 compared with 1.592 in the earlier approximation [6]. The differences are comparatively small, however. Hence, use of the 1956 treatment with its advantages of greater simplicity is vindicated for most purposes.

Table 1. Resumé of Calculations of Biphasic Equilibria in Systems of Hard Rods

	1956 Approximation			Exact Lattice Treatment		
	Neat fluid	x = 20	x = ∞	Neat fluid	x = 20	x = ∞
v_p	1	0.379	0	1	0.364	0
v_p'	1	0.541	0	1	0.498	0
v_p'/v_p	1	1.428	1.592	1	1.367	1.465
xv_p	6.70	7.58	8.28	6.42	7.28	7.89
$\bar{y}/x$	0.166	0.167	0.186	0.194	0.202	0.232

The volume fraction at incipient separation of the nematic phase lies just beyond the volume fraction v_p^* at which Z_M exhibits critical behavior with respect to the disorder index $\bar{y}$, according to the 1956 theory. The confusion of v_p^* calculated using Eq. (11) with v_p at incipience therefore has led to underestimation of the latter volume fraction according to that version of the theory. The improved theory leads to somewhat lower values of the volume fraction at incipient phase separation. Hence, the error attending misapplication of Eq. (11) is offset by inaccuracies of the original theory, with the fortuitous consequence that this equation offers a slightly better approximation for the threshold volume fraction than for v_p^*, for which latter it was originally intended. The widespread misuse of Eq. (11) is therefore well justified.

3 Binary Lyotropic Systems. Comparison of Theory with Experiment

The volume fraction v_p at which the nematic phase first appears in detectable amount is the experimentally measurable quantity most accessible for testing theory. The volume fraction v_p' in the coexisting nematic phase and the volume fraction in the system as a whole at which the isotropic phase disappears are also useful for this purpose. Measurement of these characteristics of the biphasic equilibrium usually is more difficult, however, than determination of the threshold volume fraction v_p. Both v_p and v_p' are subject to effects of polydispersity, which are discussed in the following section of this chapter. Here we examine results of experiments interpreted in the approximation that the polymer in a lyotropic system may be treated as a single component. The weight average axial ratio, obtained from the weight average molecular weight $\bar{M}_w$, appears to be an appropriate average for the purpose (cf. seq.).[4] Complete suppression of the effects of polydispersity cannot be expected by this choice, however, especially if the distribution is very broad.

Experiments on liquid crystallinity exhibited by solutions of α-helical polypeptides [10] have been especially illuminating (see Chapter 2 by Uematsu and Uematsu). Principal results are summarized and compared with theory in Table 2. Most of the measurements presented were carried out by visual observation between crossed

Table 2. Biphasic Equilibria in Solutions of α-Helical Poly(γ-Benzyl-L-Glutamate)

Mol. Wt.[a] $\times 10^{-3}$	x[b]	v_p[c] Calcd.	Expt' 1. Results			Source
			Diluent	v_p	v_p'	
220	97	0.081	m-cresol	0.078[d]		Hermans [11]
270	120	0.066	m-cresol	0.060[d]		Hermans [11]
340	150	0.053	m-cresol	0.046[d]		Hermans [11]
340	150	0.053	DMF-MeOH	0.05	0.075	Nakajima et al. [12]
790	350	0.023	DMF-MeOH	0.018	0.030	Nakajima et al. [12]
310	137	0.058	DMF-MeOH	0.05	0.06	Wee and Miller [13]
335	148	0.053	m-cresol	0.070		Kiss and Porter [14]

[a] Molecular weights are weight averages obtained from intrinsic viscosities of solutions in solvents in which PBLG adopts random coiled configurations;
[d] Axial ratios calculated using density $\varrho = 1.36$ g cm^{-3} and $l_u = 1.5$ Å; see footnote 4;
[c] Threshold volume fractions calculated according to Eq. (11);
[d] These values were determined by the viscosity method; see text

---

4 Axial ratios may be calculated from chain diameters given by

$$d = (M_u/\varrho N_A l_u)^{1/2}$$

where M_u is the molecular weight per repeat unit, ϱ is the density of the polymer, l_u is the length of the repeat unit projected on the molecular axis and N_A is the Avogadro number. Then

$$x = (M/M_u) l_u/d$$
$$= M(N_A\varrho)^{1/2} (l_u/M_u)^{3/2}$$

where M is the molecular weight (weight average).

polarizers. The results of Hermans [11], however, were obtained by identification of the threshold volume fraction v_p with the sharp peak in the viscosity measured as a function of concentration at low rates of shear. Subsequent investigations [15-17] have shown that phase separation commences somewhat before the viscosity reaches its maximum. Values of v_p obtained by the exceptionally convenient viscosity method introduced by Hermans [11] may be slightly too high on this account, perhaps as much as ten percent.

The agreement between observed and calculated values of the threshold volume fraction v_p in Table 2 is quite satisfactory. Observed values are generally somewhat smaller than those calculated, but the differences are scarcely beyond the experimental error. They may be vitiated to some extent by polydispersity, effects of which may not be suppressed entirely by use of the weight average of the chain length, or axial ratio. The ratios v'_p/v_p vary from ~1.2 according to Wee and Miller [13] to ~1.6 according to Nakajima et al. The mean is in the range predicted for a monodisperse solute; cf. seq.

It should be noted that the tendency to form a nematic phase may be markedly diminished by departures from perfect helicity of the polypeptide chain. Experiments must therefore be conducted under conditions far removed from the helix-coil transition. Also, steric purity of the monomer units must be assured. Through their effect in disrupting perpetuation of the helix, as little as one percent of D-peptide units in an L-chain (or *vice versa*) may be expected to increase v_p considerably if the degree of polymerization is large, e.g., for a degree of polymerization DP > 1000, or x > 100.

The incidence of liquid crystallinity in solutions of polymeric aramides such as poly(p-benzamide) (PBA) and poly(p-phenyleneterephthalamide) (PPDT) has been extensively investigated, both in N,N-dimethylacetamide (DMAc) containing 3–4% of LiCl and in sulfuric acid (see Chapter 3 by S. L. Papkov). Volume fractions v_p at which separation of a nematic phase occurs are plotted against the axial ratio x in Fig. 4. The solid line has been calculated according to Eq. (11). The open circles represent results of Papkov and co-workers [18, 19] for PBA in DMAc containing 4% LiCl. If the contentions of Schaefgen et al. [20] that the molecular weights determined by Papkov and co-workers [18] were vitiated by aggregation are valid then these points should be shifted to the left. Results of Kwolek et al. [21] for a large number of samples of PBA in DMAc are summarized by the lower dashed line. Results of Balbi et al. [22] on the same system are indicated by the single triangle. In sum, the results of Papkov et al. [18] indicate threshold volume fractions for PBA in DMAc that are greater than theory predicts while those of Kwolek [21] and of Balbi [22] and their co-workers fall below the theoretical line.

The upper dashed line in Figure 4 represents the experimental results of Bair, Morgan and Kilian [23] for PPDT in concentrated sulfuric acid. Their data cross the theoretical line, being slightly below it for x ≈ 100 (M ≈ 11,000) and above it for higher chain lengths. Partial flexibility of the chain (in this solvent) would account for these results inasmuch as the effective axial extension would then lag below the length of the chain at high chain lengths. In contrast to these results, Nakajima, Hirai and Hayashi [24] obtained good agreement with theory for PPDT with x = 240 and 350 in sulfuric acid containing *ca.* 15% of water, as shown by the filled circles in Fig. 4. Results for larger axial ratios (not included in Fig. 4) occur above the theoretical

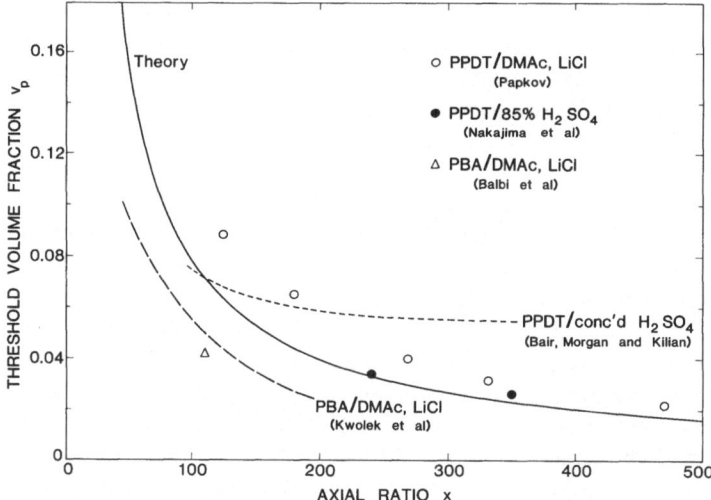

Fig. 4. Threshold volume fractions v_p for incipience of a nematic phase plotted against the axial ratio x. The solid curve is theoretical according to Eq. (11). Experimental results on PBA in DMAc are shown by the open circles (Papkov and co-workers [18]), by the long dashed line (Kwolek et al. [21]) and by the triangle (Balbi et al. [22]). The short dashed line represents measurements (Bair, Morgan and Kilian [23]) on PPDT in concentrated H_2SO_4, and the filled circles are for PPDT in 85% aqueous H_2SO_4 (Nakajima, Hirai and Hayashi [24]). Concentrations reported in w/v units have been converted to volume fractions on the basis of a density of 1.46 g cm^{-3} for the aramide polymers [21]

curve, but the deviation is smaller than for concentrated sulfuric acid. The ratios v_p'/v_p of the concentrations in the coexisting phases for the polyaramides fall in the range 1.3–1.6 [19, 23], in approximate agreement with theory for monodisperse rods. The biphasic gap appears to be little affected by the polydispersities [20] of these polymers, contrary to predictions of theory; cf. seq.

Lyotropic systems in which the nematogenic component is a poly(N-alkyl isocyanate) (PIC),

$$\left(\begin{array}{c} R \quad O \\ | \quad || \\ -N-C- \end{array} \right)_n$$

where R is a substituent of sufficient size to confer high extension on the chain, are attractive candidates for investigation owing to the favorable solubilities of PIC's in common solvents. They have been investigated extensively by Aharoni [15, 25-27]. His examinations of PHIC [26] with R = n-hexyl and POIC [27] with R = n-octyl in toluene yield threshold volume fractions v_p ca. twice those calculated from their axial ratios according to Eq. (11). For example, for PHIC with $M_w = 73,000$, for which $x \approx 100$, v_p is 0.16 according to experiments compared with a 0.08 calculated from theory. The length of the chain measures ca. 1000 Å, which is fully twice the persistence length [28]. From this it may be inferred that flexibility of the chain is sufficient to lower the effective axial ratio significantly for chains of the quoted length.

Shear-degraded DNA's provide rodlike particles of convenient length for observations on liquid crystallinity, according to the work of Brian, Frisch and Lerman [29]. For a sample with *number average* molecular weight $1.1_3 \times 10^5$ a nematic phase was observed to appear at $v_p = 0.24$ in aqueous 2M NaCl. Measurements in 1M salt solution yielded $v_p = 0.26$, in close agreement with this result. The effective axial ratio is $x = 24.8$ as calculated from the length of the DNA double helix and its diameter deduced from the osmotic equation of state. (The effective diameter, 23.4 Å in 2M salt, thus obtained is in close agreement with results of X-ray diffraction on DNA fibers). On this basis we calculate $v_p = 0.30$ according to Eq. (11), which is only marginally greater than the values observed.

Although the polydispersity of the shear-degraded DNA is indicated to be low [29], correction of x to the more appropriate weight average axial ratio would lower the calculated value of v_p and, hence, narrow the discrepancy. The agreement with theory appears, therefore, to be well within the limits of experimental uncertainties. Brian et al. [29] also determined the concentration of DNA in the coexisting anisotropic phase. Their results yield 1.4–1.5 for the ratio of the concentrations in the two phases, in good agreement with the lattice theory.

A further aspect of lyotropic solutions of extended-chain polymers that merits investigation is the degree of order in the nematic phase. The lattice theory [6, 7] makes explicit predictions in this regard, both in terms of the disorder index $\bar{y}$ and of the familiar order parameter $s = (3\langle \cos^2 \psi \rangle - 1)/2$. Although this property of nematic phases of low molecular nematogens has received much attention, it has been largely ignored in investigations of liquid crystallinity in polymers.

4 Multicomponent Systems

4.1 Polydisperse Rods

The lattice theory is readily adapted to the treatment of a system of homologous rodlike molecules differing in length and, correspondingly, in axial ratio [9, 30, 31]. The approximate theory [6] is especially tractable for this purpose [9]. One immediately deduces from the theory developed on this basis that all of the homologous species acquire the same value of $\bar{y}$ at orientational equilibrium, with the exception, of course, of any species for which x is less than the value $\bar{y}$ common to all larger species. The degree of disorientation as measured by $\bar{y}/x$, or by $\langle \sin \psi \rangle$ (see Eq. (A-6)) in Appendix-A), decreases with x, as should be expected, but in a manner such as to engage every species in the same average number $\bar{y}$ of lattice rows, regardless of their axial ratios x (for $x \geq \bar{y}$). The universality of the parameter $\bar{y}$, which follows from the 1956 theory without further postulates or approximations, greatly simplifies theoretical treatment of polydisperse systems.

Numerical calculations have been carried out for (i) ternary systems consisting of rods of two lengths x_a and x_b and a diluent with $x = 1$ [32], (ii) solutions of poly-disperse rods having a "most probable" distribution [33], (iii) a Poisson distribution of rods in solution [34], and (iv) various Gaussian distributions of rods in a diluent [35]. In all cases longer rods are preferentially partitioned into the nematic phase. For $x_a = 2x_b$ in case (i) [32] the ratio of the concentration of either of the species in one of

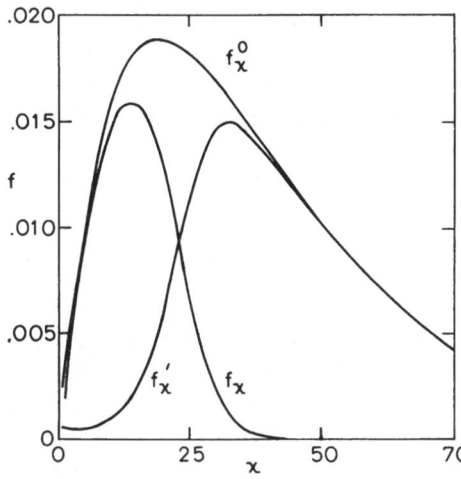

Fig. 5. Weight fractions f_x of species of a "most probable" distribution of rods apportioned between two phases at an overall dilution such that the volumes of the two phases are equal. The uppermost curve represents the f_x^0 for the unseparated distribution; $\bar{x}_n^0 = 20$. The lower curves show the distributions f_x and f_x' calculated according to theory for the isotropic and nematic phases, respectively, in which $\bar{x}_n = 10.4$ and $\bar{x}_n' = 36.5$ (From Flory and Frost [33])

the phases to its concentration in the other phase may be as great as four. Marked fractionation between the two phases is predicted also in case (ii) [33]. Calculations for a most probable distribution with $\bar{x}_n = 20$ and $\bar{x}_w = 39$ are shown in Fig. 5, the total volume having been so chosen as to render the volumes of the two phases equal. The lower curves represent the distributions of weight fractions of the various species in the respective phases; the upper curve shows the unseparated distribution for the whole polymer. The selectivity with which the components are partitioned between the two phases is apparent; higher members occur almost exclusively in the nematic phase, the lower ones preferentially in the isotropic phase.

Fractionation is less pronounced for the much narrower Poisson distribution [34]. For a Gaussian distribution [35] it varies with the breadth assigned to that distribution, as expected.

Concomitantly with the partitioning of solute species between the coexisting phases, broadening of the biphasic gap (measured by the concentration difference between the two phases) is predicted [32-35], the more so the greater the disparity in the sizes of the species prominently present.

Experiments confirm these predictions qualitatively but not quantitatively. We have mentioned above that the ratios, v_p'/v_p, of the volume fractions in the coexisting phases for unfractionated polypeptides [12,13], polyaramides [19,22,23,36] and polyisocyanates [15,26,27] are little greater than the ratios calculated for monodisperse rods, namely, 1.46 in the limit $x \to \infty$. According to the theoretical calculations this ratio for a "most probable" distribution should be in the range of 2–3, depending on the average chain length and the ratio of volumes of the two phases. Bair, Morgan and Kilian [23] found $v_p'/v_p \approx 1.8$ for unfractionated poly(chloro-p-phenyleneterephthalamide) (ClPPDT) in DMAc, LiCl. Other results are mostly in the range 1.4–1.6, however. It should be observed that separation of the two phases is difficult and each of them may be contaminated with the other. Some of the experiments may be subject to error on this account.

The predicted fractionation of species between the isotropic and nematic phases is universally confirmed [15,22,23,25,27,36], but the selectivity is generally somewhat less than theory predicts. However, Bair et al. [23] found the ratio η'_{inh}/η_{inh} of inherent viscosities in the coexisting phases in the system C1PPDT/DMAc, LiCl to be ca. 3.3, in tolerable agreement with theory [33] for a most probable distribution. The largest value of the ratio $\bar{x}'/\bar{x}$ found for PBA/DMAc, LiCl by Balbi et al. [22] is 2.5, which is less than theory predicts. Aharoni [27] reports substantial fractionation when two polyisocyanates (POIC) differing in chain length are dissolved in toluene and then allowed to phase-separate.

Recent results [37, 38] lead to the conclusion that compositional equilibration may not readily be attained in the biphasic separation of mixtures consisting of two or more nematogenic components. Theory clearly indicates that orientational order supersedes composition in its importance in differentiating the coexisting phases. Gradients of the chemical potentials are small. Whereas orientational ordering of the molecules may occur rapidly, partitioning of components between phases depends on diffusive transport, a much slower process once the phases acquire macroscopic dimensions. For this reason, the degree of partitioning of species between phases predicted by theory for the system at equilibrium may not be realized in experiments as normally conducted.

4.2 Mixtures of Rods and Random Coils

Application of the lattice scheme to a system consisting of rodlike molecules, random coils and a solvent is straightforward [39]. The free energy of the nematic phase would be increased markedly by the presence of a substantial proportion of the random coil.

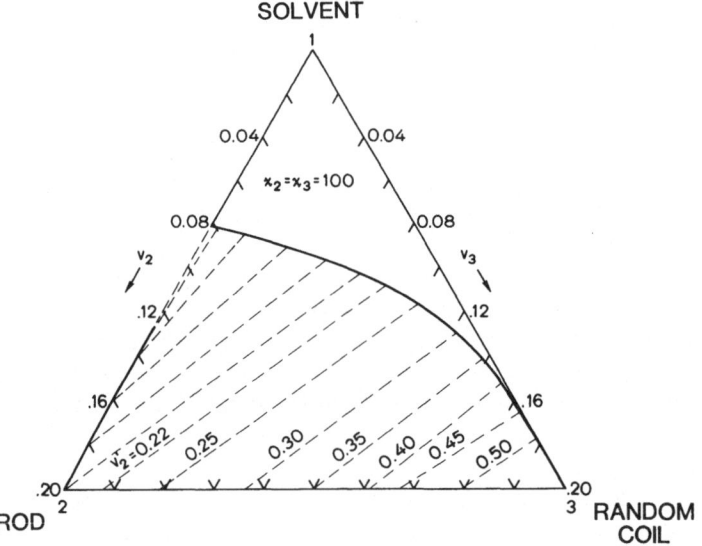

Fig. 6. Phase diagram calculated for the ternary system: solvent (1)/rigid rod (2)/random coil (3) with $x_1 = 1$ and $x_2 = x_3 = 100$. Numerals on the tie lines that extend below the lower boundary of the diagram denote v'_2 for the conjugate anisotropic phase the binodal for which lies virtually on the $1 - 2$ axis (From Ref. [39])

The latter component consequently is predicted to be virtually excluded from this phase. This striking deduction from theory follows solely from the "steric" requirements of the two kinds of macromolecules; no other interactions between them are required.

The phase diagram calculated [39] for such a system comprising a solvent (1) with $x_1 = 1$ and rods (2) and coils (3), with $x_2 = x_3 = 100$, is shown in Fig. 6. The curve across the upper portion of the diagram is the binodal representing the locus of compositions for the coexisting isotropic phase. The binodal for the nematic phase virtually coincides with the axis for components 1 and 2; the concentration of random coils (3) in this phase is calculated to be immeasurably small. Dashed lines in Fig. 6 denote tie-lines; for those that extend below the range of the diagram, the volume fraction v_p' at the terminus is quoted on the figure. The isotropic phase tolerates an appreciable proportion of the rodlike component if the concentration of the random coil therein is not too great. As the latter concentration is increased much beyond the threshold concentration in the binary system 1–2, however, the proportion of the rodlike component (2) in this phase becomes negligibly small. Except within a limited range of composition, therefore, the two polymeric components may be regarded as mutually incompatible, according to theory.

Aharoni [40] has reported experiments on the system tetrachloroethane/polyisocyanate (POIC)/polystyrene that confirm the foregoing predictions. The nematic phase separates at compositions in the range predicted; polystyrene does not occur in detectable quantities in this phase. The binodal for the isotropic phase resembles the calculated curve shown in Fig. 6.

Similar results have been obtained by Bianchi, Ciferri and Tealdi [41] for the system: DMAc, 3% LiCl/PBA/X-500; the lattermost designation denotes the polymeric terephthalamide of p-aminobenzhydrazide which occurs as a random coil in solution. Their experimental results are shown in Fig. 7. In confirmation of Aharoni's results [40], and of theory [39] as well, the "X-500" component is not detectable in the nematic phase. The close resemblance of Figure 7 to the theoretical diagram shown in Figure 6 is evident.

Recently Hwang, Wiff and Benner [42] have determined the phase diagram for the ternary system: methanesulfonic acid/poly(phenylene benzothiazole) (PBT)/poly-

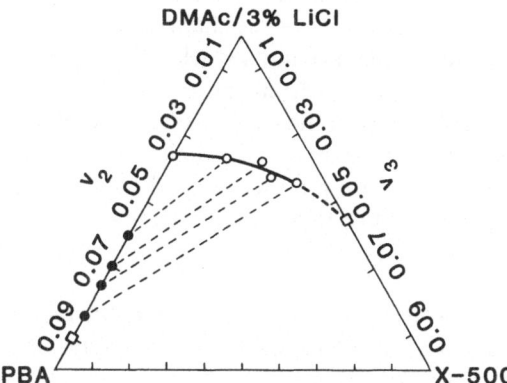

Fig. 7. Phase diagram observed for the ternary system indicated, PBA being the rigid rodlike component and "X-500" the random coil. See text (From Bianchi, Ciferri and Tealdi [41])

(-2,5(6)benzimidazole). Molecules of the second component, PBT, approach the ultimate in rodlike rigidity. The third component is a non-mesogenic random coil. The phase diagram determined by Hwang, Wiff and Benner [42] for this system bears a close resemblance to Fig. 6.

Reflecting on the unique features for these ternary systems, we observe that, in rejecting the random coil, the nematic phase manifests a principal characteristic of crystals in general, thus validating the term "liquid *crystal*". It will be apparent, however, that the presence of the random coil in the system affects the composition of the anisotropic phase substantially, its absence therein notwithstanding. This is a consequence of the abundance of random coiled component in the coexisting isotropic phase; it markedly increases the chemical potential of the rodlike component in that phase and, concomitantly, decreases the chemical potential of the diluent. The concentration of the rodlike component in the anisotropic phase consequently is caused to increase.

5 Semirigid Chains

The examples cited above are illustrative of polymers whose structure and conformation enforce a high degree of extension with only minor departures of the direction of the chain axis over the length of the molecule. This degree of rigidity is not required for manifestation of liquid crystallinity. Semirigid, or semiflexible, chains which, though highly extended compared to familiar random coiling polymers, nevertheless possess a significant measure of flexibility may impart liquid crystallinity to their melts and solutions [3, 4]. The concentration required for separation of a nematic phase from solution is, as should be expected, greater than for rigid rods of the same molecular length. It is obviously important to inquire into the effect of departures from full rigidity on the isotropic-nematic transition.

Such departures may occur in diverse ways in various semirigid chains. At one extreme, we have chains appropriately represented by the wormlike model in which each unit deviates slightly from the direction of its predecessor, successive deviations being uncorrelated in magnitude and direction. Even in the examples of "rigid" chains considered in the preceding Section, the cumulative effects of small deviations of this kind cause the chain to acquire the character of a random coil at very high chain lengths, beyond the ranges covered in the experiments cited above.

At the other extreme are chains in which the direction of the chain axis undergoes abrupt changes at occasional points along the chain. The units at which these abrupt "bends" occur may differ chemically (or structurally) from the principal units of the chain, as in a copolymer, or they may differ only in their conformation. Cellulose and cellulose derivatives in which the dominant source of flexibility arises from pseudorotation of an occasional sugar ring [43], causing the O—C and C—O bonds pendant to the aberrant ring to adopt transverse directions, are illustrative. Polymer chains consisting of alternating sequences of rigid and flexible units provide further examples which may exhibit liquid crystallinity if the rigid sequences are sufficiently long [44, 45]. Polymers of these types are not well represented by the wormlike model.

Inasmuch as semiflexibility, or semirigidity, can be realized in diverse ways, it is scarcely to be expected that a single theory will comprehend all polymers of this kind.

A number of theories have been proposed [46,47]5. A particularly simple model that provides intuitive insight into the tendency of a semirigid chain to induce formation of a nematic phase is the Kuhn chain consisting of bonds of appropriate length connected by flexible joints. This obviously artificial model offers the distinct advantage of being susceptible to treatment within the framework of the lattice scheme presented above. Theoretical predictions [48] on biphasic equilibria, derived in the 1956 approximation, differ negligibly from those for independent rods having the same axial ratio as the Kuhn segments. Thus, the effect of joining these segments linearly by connections which allow complete flexibility, directions of successive segments being uncorrelated, is quite negligible for most purposes. This result can be understood [48] from the realization that characteristics of a nematic phase are dominated by interactions of higher order than those comprehended in the first virial coefficient, which is the quantity exclusively affected by connecting the segments.

The foregoing result suggests that the theory for rigid rods may be adapted to semirigid chains by the simple device of replacing the molecular axial ratio x in the theory above by the axial ratio x_K of the Kuhn segment. This latter quantity can be calculated from the mean-square end-to-end length $\langle r^2 \rangle_0$ of the unperturbed, random chain and its length L at full extension. Thus, the relations $n_K l_K^2 = \langle r^2 \rangle_0$ and $L = n_K l_K$, where n_K is the number of segments in the model chain and l_K is the length of one of them, serve to specify these parameters. Moreover, $L = n l_u$ where n and l_u are, respectively, the number of repeating units in the real chain and the projected length of one of them on the axis of the extended chain. It follows that

$$l_K = \langle r^2 \rangle_0 / n l_u$$

With the chain diameter given by

$$d = (M_u / N_A \varrho l_u)^{1/2}$$

(see footnote 4, one obtains)

$$x_K = l_K / d = (\langle r^2 \rangle_0 / n)(\varrho N_A / l_u M_u)^{1/2} \tag{13}$$

where ϱ is the density of the polymer and M_u is the molecular weight of the repeating unit.

---

5 The semiflexible lattice chain introduced in Ref. [46] has been widely applied to mesogenic polymers, and to other polymers as well. In the opinion of the author, many of these applications go beyond the limitations of the model; it was originally introduced for illustration of the effects to be expected from low degrees of flexibility in long chain molecules, and not as a model for representing real chains generally. According to this model, perpetuation of the chain along an axis defined by preceding segments represents the configuration of minimum energy. Departures may occur at angles of $\pi/2$ from this direction; they entail an energy $\varepsilon > 0$. This difference in energy favoring the rectilinear form is presumed to determine the temperature coefficient of the chain dimensions. The relationship thus prescribed is usually inaccurate at best; not infrequently the observed temperature coefficient is positive instead of negative as the model would require. With specific reference to nematogenic polymers, the model does not lend itself to allowance for the persistence of correlations of successive rectilinear sequences with the domain axis, unless perfect alignment is assumed in the anisotropic phase [46]. Finite degrees of disorder in this phase are difficult to take into account.

5.1 Cellulose and its Derivatives

Threshold volume fractions observed for cellulose acetate (CA and CTA), ethyl cellulose (EC) and hydroxypropyl cellulose (HPC), each in various solvents, are presented in Table 3. The results depend to some extent on the solvent [49]. The data included are not exhaustive. Other cellulose esters exhibit mesomorphic behavior [51]. Chanzy et al. [54] observed mesomorphic behavior in solutions of cellulose itself when dissolved in N-methylmorpholine N-oxide containing water at concentrations of cellulose in the range 20–55% w/v, depending on the temperature, the water content of the solvent and the degree of polymerization of the cellulose. Solutions of cellulose in mixtures of trifluoroacetic acid with 1,2-dichloroethane or with chloroform are likewise lyotropic at concentrations of 20% (w/v) and above according to Patel and Gilbert [55].

The cellulose derivatives chosen for inclusion in Table 3 are those for which the required data for comparison with the theory as outlined above are available. For cellulose acetate, $\langle r^2 \rangle_0/n \approx 1100 \pm 200$ Å^2 in the limit $n \to \infty$ [56, 6]. Taking

Table 3. Incipience of Cholesteric Phase Separation from Solutions of Cellulose Derivatives

Derivative[a]	Solvent	Threshold volume fraction v_p
CA	acetone	0.33[b]
CA	$HCON(CH_3)_2$	0.37[b]
CA	dioxane	0.35[b]
CA	m-cresol	0.28[b]
CA	CF_3COOH	0.25[c]
CTA	$Cl_2CHCOOH$	0.35[d]
EC	CH_3COOH	0.39[d]
EC	$Cl_2CHCOOH$	0.28[d]
HPC	H_2O	0.39[d], 0.37[e], 0.32[f]
HPC	CH_3OH	0.33[e]
HPC	C_2H_5OH	0.39[d], 0.37[e]
HPC	CH_3COOH	0.27[d]
HPC	$CH_3CON(CH_3)_2$	0.35[d], 0.32[f]

[a] CA = cellulose acetate, degree of substitution 2.45;
   CTA = cellulose triacetate;
   EC = ethyl cellulose;
   HPC = hydroxypropyl cellulose;
[b] Aharoni [49];
[c] Dayan, Maissa, Vellutini and Sixou [50];
[d] Bheda, Fellers and White [51];
[e] Werbowyj and Gray [52];
[f] Conio, Bianchi, Ciferri, Tealdi and Aden [53]

---

6 Tanner and Berry [56] obtained $\langle r^2 \rangle_0/n = 1080$ Å^2 for a secondary acetate (CA) of degree of substitution 2.45 dissolved in trifluoroacetic acid or in a mixture of methylene chloride and methanol. They found larger values of 1350 or greater for the triacetate (CTA); Kamide, Miyazaki and Abe [57], on the other hand, report lower values of ca. 600 for CTA. The disparity may reflect the difficulties caused by aggregation of CTA [56]. Hence, we adopt the same value for CTA as for CA, on the plausible grounds that they should be similar in this respect

$\varrho = 1.3 \text{ g cm}^{-3}$, $l_u = 5.2 \text{ Å}$ and $M_u = 265$ (for CA), one obtains $x_K = 26$ according to Eq. (13), and, from Eq. (11), $v_p = 0.284$ at incipience of a nematic (or cholesteric) phase. The experimental values in Table 3 are generally somewhat greater than this result, but the differences are not beyond the uncertainty in the characteristic chain dimension, here represented by $\langle r^2 \rangle_0/n$. Inasmuch as this ratio depends on the solvent, as is well known, dependence of the threshold v_p on the solvent [49] is to be expected. Available data do not permit a definitive correlation, however, between v_p at incipience of the mesomorphic phase and the influence of the solvent on chain flexibility.

Navard et al. [58] have investigated the dependence of v_p for CA on temperature in trifluoroactic acid. Its increase with temperature is consistent, approximately, with the decrease in characteristic dimensions with temperature [59].

According to Wirick and Waldman [60] $\langle r^2 \rangle_0/n \approx 1000 \text{ Å}^2$ for cellulose ethers, hence, $l_K \approx 190 \text{ Å}$. Werbowj and Gray [52] conclude on the basis of examination of available data that $l_K = 170 \pm 40 \text{ Å}$. Taking an intermediate value of $l_K = 180 \text{ Å}$, we obtain $x_K = 23$ and $v_p = 0.32$ for EC ($\varrho = 1.3 \text{ g cm}^{-3}$), and $x_K = 19$ and $v_p = 0.38$ for HPC (trisubstituted; $\varrho = 1.2 \text{ g cm}^{-3}$; see Ref. [52]). Experimental results quoted in Table 3 for both EC and HPC are consistent with these calculations; values observed for HPC are generally somewhat lower than the calculated value, but the differences are small.

For EC, Bheda, Fellers and White [51] find $v_p'/v_p = 1.43$ in dichloroacetic acid and 1.18 in acetic acid. For HPC, they report $v_p'/v_p = 1.29$ to 1.37 in four different solvents. These results are compatible with the theory for monodisperse rods. It is to be noted that polydispersity of segment lengths should be unimportant if the chains are uniform in structure and composition or, more generally, if dissipation of directional persistence over the span of a Kuhn segment is the result of a number of restricted changes in chain direction. Hence, corrections for polydispersity should not be required for chains conforming to the representation here adopted.

5.2 Other Polysaccharides

Other polysaccharides whose solutions yield liquid crystalline phases include the bacterial polysaccharide xylan [61] and the extracellular fungal polysaccharide produced by *schizophyllum commune* [62].

The former is a β-1,4-D-glucan with a side chain consisting of three sugar units attached to every second residue of the backbone. Its extended form [61], possibly a double helix, has a persistence length of *ca.* 500 Å [63]. It is a polyelectrolyte in aqueous solutions. The latter polysaccharide is a β-1,3-D-glucan in which every third unit bears a single glucose residue as a side chain. It is uncharged. In aqueous solution it forms a triple helix [64, 65] having a persistence length of $1800 \pm 300 \text{ Å}$ with a pitch per backbone residue of 3.0 Å along the helix axis [64, 65]. Taking the density to be 1.6 g cm^{-3} [65], one obtains $d = 15 \text{ Å}$ for the chain diameter, which is smaller than estimates of 20—32 Å from hydrodynamic measurements [66] and of 16–28 Å from light scattering [67]. The result based on the density may be presumed to be the most reliable. On this basis, $x_K = 240$ and the threshold volume fraction calculated from Eq. (11) is 0.033. Using a schizophyllan having a molecular weight of 4×10^6 (which is *ca.* five times the molecular weight corresponding to the Kuhn length) Van, Norisuye and Teramoto [62] determined the weight concentrations in the isotropic and cholesteric phases coexisting at equilibrium in water at 25 °C. Conversion of their results to volume fractions according to the quoted density gives $v_p = 0.064$ and $v_p' = 0.085$. The ratio is in satisfactory agreement with theory but the magnitudes are about twice those calculated. The departures of the schizophyllan helix from strict rigidity that are responsible for its finite persistence doubtless comprise many very small deviations, in conformity with the wormlike model chain. This contrasts with the relatively large, but less frequent, changes of direction believed to be operative in the single-stranded cellulosic chains. The results for the schizophyllan triple helix suggest that the Kuhn segment required to replicate a wormlike chain may not be equivalent, quantitatively, to a free rod in regard to biphasic equilibrium. Experimental results on other semirigid chains, or compound helices, obviously are needed [7].

5.3 Polymers Comprising Both Rigid and Flexible Units or Sequences

Polymers consisting of rigid and flexible units, or sequences of units, in alternating succession have attracted widespread interest owing to their propensity to exhibit mesomorphic behavior in the melt, and because of the similarties between their rigid sequences and low molecular nematogens [44,45,68]. They may be smectic or nematic. Their melts are thermotropic, and in some instances their solutions exhibit lyotropic behavior. Typical examples are polymers comprising units of the type

$$\left[CO-O-\langle\bigcirc\rangle-C(CH_3)=CH-\langle\bigcirc\rangle-O-CO-(CH_2)_n \right]$$

investigated by Roviello and Sirigu [45]. Members of the series with $n = 6-12$ melt to nematic fluids. Nematic-isotropic transition temperatures for the polymers with n even decrease from 300 °C for $n = 6$ to 190 °C for $n = 12$. Homologs with n odd exhibit lower transition temperatures than the adjacent even members of the series. Oscillation of transition temperatures between successive even and odd members of the series persists even for $n > 10$, where it is on the order of 10–20 °C. This oscillation is indicative of correlations between the consecutive rigid sequences propagated through a flexible chain comprising as many as twelve bonds [69]. A similar pattern of behavior is observed for other alternating polymers of this kind [44,68].

If the random sequence between successive rigid members in the chain should be of sufficient length and flexibility to eliminate correlations between these members, then the lattice methods described and discussed above are readily applicable [31]. Theory indicates a resemblance to the mixtures of rigid rods and random coils discussed above. In the polymers under consideration, however, the random and the rigid components are combined in the same molecule. Hence, they are prevented from segregating in the respective coexisting phases of a biphasic system, the thermodynamic driving forces favoring such separation notwithstanding. These "alternating copolymers" may be expected to partake of some of the characteristics of the Kuhn

---

7 It will be evident that theory could be brought into agreement with experiment through arbitrary choice of an equivalent segment equal in length to the persistence of the schizophyllan triple helix. Independent justification for this choice is not apparent, however.

model chain having volumeless flexible joints on the one hand and of mixtures of rods and random coils on the other. Application of the lattice model, under the assumption that successive rigid sequences are uncorrelated, leads to the prediction that a small fraction of random-coil units combined with rigid members should suffice to affect markedly the phase diagram for a lyotropic system [31]. The presence of such units should increase the disorientation (y) in anisotropic phase. The biphasic gap may be narrowed and the partitioning of large and small species between the nematic and isotropic phases should be diminished, according to theory [31].

Even for a "flexible" sequence consisting of ten or so methylene groups, the correlation between the adjoining rodlike sequences persists to an appreciable degree according to the results discussed. This observation is affirmed by conformational studies conducted by Abe [69]. Hence, the assumption that successive rigid sequences in a given chain may adopt mutually independent orientations is inaccurate at best and may be quite untenable. Incorporation of these correlations in the theory presents serious difficulties owing to the fact that each sequence is subject to constraints both by adjoining members in the same chain and by neighboring sequences of surrounding chains. A satisfactory theory that takes account of the simultaneous effect of both constraints has yet to be developed.

Mesomorphic behavior has been observed in a number of random copolymers consisting of two kinds of units, one rigid and rodlike and the other "flexible", in the sense of offering a variety of conformations and, hence, capable of disrupting perpetuation of the axis of the rigid unit. The p-hydroxybenzoate (PHB) unit is prototypal of a rigid unit; it has often been incorporated with ethylene terephthalate (ET) as the flexible unit present as the minor constituent (0.3–0.4 mole fraction). [68, 70] If the proportion of the flexible unit is sufficient to reduce the crystalline melting point below the thermal decomposition temperature, the copolymer may be expected to melt to a mesomorphic fluid, provided that the proportion of the rigid unit suffices to generate long sequences of that unit, consecutively linked, in adequate abundance. If these sequences should be selectively sequestered into the nematic phase, intervening portions of the chains that include many flexible units being relegated to the isotropic phase, then the nematic domains would necessarily be very small. This follows from the intermingling of sequences of diverse composition in the same chain. The resulting morphology would resemble that of a semicrystalline copolymer in which the co-unit is excluded from the crystalline phase.

While the possibility of this dispersion of micro-domains of the nematic phase in an isotropic phase cannot be dismissed, concrete evidence for morphologies of this kind in nematogenic copolymers is not prominently in evidence. The longer sequences of rigid units undoubtedly are responsible for promotion of liquid crystallinity but, as theory suggests [31], they appear to be uniformly dispersed [68]. Sequences of units that include many flexible members and hence are not rodlike may assume a role analogous to that of the solvent in a lyotropic system [31]. The nematic copolymer should, on this basis, consist of a single phase.

The "continuous-phase" morphology is firmly established for copolymers of HBA and ET, mentioned above, by the recent work of Windle and collaborators [71, 72]. They observed birefringent domains in sections of oriented pellets obtained by extrusion of these copolymers. The combined techniques of X-ray diffraction and optical birefringence revealed biaxial order within the domains, instead of the uniaxial order usually found in nematics. The origin of the biaxial order is unclear. However, the perpetuation of that order throughout domains measuring several μm in diameter is unambiguous evidence for prevalénce of homogeneity as opposed to separate micro-domains, whose axes would be uncorrelated (apart from the uniaxial order imparted by the process of extrusion).

Organization at the molecular level in nematic copolymers may depart in an important respect from that occurring in nematic phases (typically lyotropic solutions) of rigid-chain homopolymers. The virtual elimination of entanglements that is implicit in the creation of nematic domains much larger than molecular dimensions in the case of rigid homopolymers may not, in general, be realized in random copolymers that comprise flexible as well as rigid units. The presence of sequences containing many flexible units allows intertwining of the copolymeric chains and hence promotes a kind of disorder not present in nematic homopolymers. This fundamental difference in morphology should be reflected in inferior physical properties of the materials formed from such mesogenic copolymers.

5.4 Conformational Changes Coupled with the Isotropic-Nematic Transition

Throughout the preceding discussions of homopolymers we have assumed that the conformations of the polymer chains are the same in both the nematic and isotropic states. It is readily apparent that this assumption must fail if the nematogenic chain molecules are able to adopt a variety of configurations, some more extended and rodlike than others. The preferential partitioning of longer species of a mixture of rods into the nematic phase of a biphasic system, predicted by theory and confirmed by experiments, and, similarly, the exclusion of random chains from the nematic phase imply unambiguously that if alternative configurations are accessible to the molecules, then the averaged conformations in the two phases must differ. Beyond these qualitative inferences, theory has yet to be developed that will shed light on this important issue as it pertains to semirigid chains in general. Certain special cases have been treated however [73, 74].

An extreme case that has received attention is that of a cooperative conformational transition examplified by the coil-helix (c → h) transition that occurs in poly-α-aminoacids. Whereas the polypeptide chain is quite flexible when it exists as a random coil, the rigid helical form may bring about formation of a liquid crystalline phase, as discussed above, if its concentration is sufficient. The conformational transition and the phase transition may therefore be coupled. The helix-coil transition may then acquire the character of a first-order phase transition, owing to generation of the liquid crystalline phase.

This conclusion was reached, tentatively, by Frenkel, Shaltyko and Elyashevich [75]. A phenomenological analysis presented by Pincus and de Gennes [76] predicted a first-order phase transition even in the absence of cooperativity in the conformational transition. These authors relied on the Maier-Saupe theory for representation of the interactions between rodlike particles. Orientation-dependent interactions of this type are attenuated by dilution in lyotropic systems generally. In the case of α-helical polypeptides they should be negligible owing to the small anisotropy of the polariz-ability of the peptide unit (cf. seq.). Moreover, the universally important "steric" interactions between the helices, regarded as hard rods, are not included in the Maier-

Saupe theory. Hence, the basis for the prediction offered by Pincus and de Gennes leaves it open to question.

A theoretical treatment has recently been carried out by the author in collaboration with Matheson [73] along the lines discussed above with appeal only to the spatial requirements of hard rods as represented in the lattice model, orientation-dependent interactions being appropriately ignored. The two transitions, one conformational and the other a cooperative intermolecular transition, are found to be mutually affected; each promotes the other as expected. The coil-helix conformational transition is markedly sharpened so that it becomes virtually discrete, and hence may be represented as a transition of first-order. These deductions follow from the steric interactions of hard rods alone; intermolecular attractive forces, either orientation-dependent or isotropic, are not required.

The high degree of cooperativity that is characteristic of the helix-coil transition is not a prerequisite for coupling of a conformational change with the isotropic-nematic transition, although it may serve to accentuate the effect. Significant effects may be expected in other circumstances.

6 "Soft" Intermolecular Interactions

The lattice theory and its various ramifications discussed above pertain to "hard" rods devoid of interactions other than the short-range repulsions that preclude intrusion of one rod on the space occupied by another. Theoretical deductions presented thus far stem from geometrical aspects of the particles or molecules, that is, from their axial ratios in the case of rigid rods and from the persistences of rigid sequences in the case of semirigid chains of the various kinds discussed above. We now consider the effects of intermolecular attractive forces between the molecules, bearing in mind that in a solution these must be assessed relative to the similar forces between the other pairs of species, i.e., solvent-solute and solvent-solvent pairs. The forces of interest fall into two categories: isotropic forces of the kind commonly treated in theories of solutions, and the anisotropic, or orientation-dependent, interactions that may contribute significantly to the stability of a nematic phase as was pointed out in the Introduction. We first discuss the effects of the more familiar interactions of the former kind.

6.1 Isotropic Interactions

Interactions of this kind are easily incorporated in the theory by adding a term of the van Laar form to the free energy of mixing, with corresponding terms in the chemical potentials. For a binary mixture the terms to be included in the reduced chemical potentials for the solvent and polymer, respectively, are χv_p^2 and $x\chi v_s^2$ (see Eqs. (C-3)–(C-6) of Appendix-C), χ being the familiar interaction parameter. These terms have little effect on the biphasic equilibrium if $\chi < 0$, i.e., if the net exchange free energy involved in the elimination of contacts between solute segments is negative [6]; the interaction between the solute molecules is effectively repulsive in this situation. If, on the other hand, the interaction between solute segments is attractive, as is the case for $\chi > 0$, the biphasic region is abruptly widened when χ

exceeds a fairly small positive value. The anisotropic phase having a concentration not much greater than that of the coexisting isotropic phase should then be supplanted by a dense, highly ordered phase, according to theory. If, additionally, the axial ratio x exceeds ca. 50 new features appear [6, 13]. Specifically, a second biphasic equilibrium energes between two anisotropic phases within a restricted range of χ.

Predictions of theory [6] for rods with axial ratio x = 100 are shown by the curves in Fig. 8. Here χ is plotted as ordinate against the volume fractions v_p and v_p' in the coexisting phases; the ordinate may alternatively, be regarded as an (inverse) measure of temperature. The narrow biphasic gap is little affected by the interactions for negative values of χ, as was noted above. If, however, χ is positive, a critical point emerges at $\chi = 0.055$. For values of χ immediately above this critical limit, the shallow concave curve delineates the loci of coexisting anisotropic phases, these being in addition to the isotropic and nematic phases of lower concentration within the narrow biphasic gap on the left. At $\chi = 0.070$ the compositions of two of the phases, one from each of the respective pairs, reach the same value. Three phases coexist at this triple point.

The general features of this phase diagram have been well confirmed by experiments, especially those of Nakajima [12, 24] and of Miller [13] and their co-workers. The emergence of a concentrated, well-ordered anisotropic phase when χ exceeds a small positive value (that depends on the axial ratio x) is readily understood as the consequence of interactions that are effectively attractive between the rodlike particles. The dense anisotropic phase may be regarded as the prototype of a quasi-crystalline state with uniaxial order only. Comprehension of the three-dimensional order characteristic of the crystalline state is, of course, beyond the scope of the model, which does not

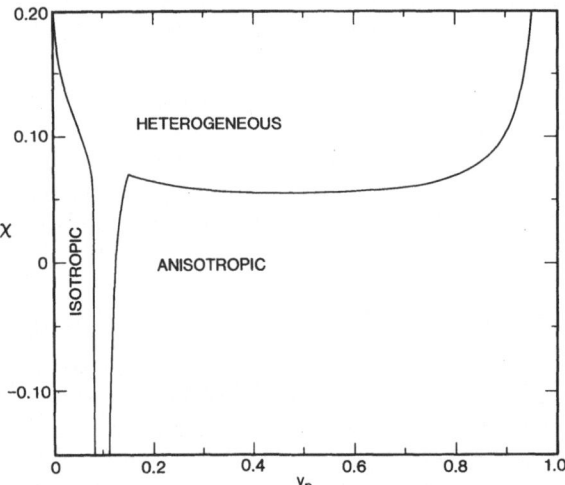

Fig. 8. Compositions in volume fractions of coexisting phases for rods of axial ratio x = 100 subject to interactions denoted by the parameter χ. The binodal for isotropic phases is on the left; that for anisotropic phases is on the right. The minimum of the shallow concave branch of the latter binodal is a critical point marking the emergence of two additional anisotropic phases. The cusp marks a triple point where three phases coexist. Calculations carried out according to the 1956 theory; see Ref. [6]

take account of specific interactions such as are required to establish longitudinal ordering of the rods.

6.2 Orientation-Dependent Interactions

The London dispersion forces between isotropic molecules or segments depend on the product of their polarizabilities α_1 and α_2. In like manner the difference in energy between mutually aligned neighbors and randomly disoriented ones depends on the product of the anisotropies of their polarizabilities. If the polarizabilities are cylindrically symmetric with respect to the long molecular axis, or if they can be so approximated, then for identical interacting segments or molecules the lowering of their mutual energy is proportional to $(\Delta\alpha)^2$, where $\Delta\alpha$ is the difference between the polarizabilities parallel and perpendicular to this axis [5, 77), 8].

It is important to observe that the relevant interaction energy is that between contacting pairs of segments of neighboring molecules and not the energy for a pair of molecules, each considered as a whole [77)]. Obviously, the interaction between all of the segments of a chosen pair of rodlike molecules must increase with improvement of their mutual alignment owing to the greater number of segments in contact. This increase in interaction occurs regardless of the anisotropy of the intermolecular forces; the anisotropy indicated by $\Delta\alpha \neq 0$ is not required. The enhancement of contacts between a given pair of molecules as they become mutually aligned is compensated by the concomitant elimination of contacts with other molecules [77)]. The compensation is exact at constant volume; hence, the net change of energy on this score should be null if the incident dispersion forces are isotropic, i.e., if $\Delta\alpha = 0$. (Since the volume change at the nematic-isotropic transition is very small, a contribution to the energy from dilation will be correspondingly small.) If, however, the anisotropy of the polarizability is appreciable, then the segment-segment interaction energy is rendered orientation-dependent and the intermolecular energy of the system is lowered by molecular alignment. This applies to the preponderance of low molecular nematogens, which usually contain the highly anisotropic p-phenylene group and other groups such as

$$-N=N- \qquad -C\equiv C- \qquad \text{and} \qquad -C\equiv N$$

with larger polarizabilities along their bond axes than perpendicular thereto.

A segment inclined at an angle ψ with respect to the domain axis experiences an energy [5, 77)]

$$\varepsilon(\psi) = -kT^* v_p s \left(1 - \frac{3}{2}\sin^2\psi\right) \tag{14}$$

---

8 Orientational energy from this source provides the basis for the Maier and Saupe theory [5)]; asymmetry of molecular shape and its influence on molecular packing in the fluid are not taken into account. Consequently, in order to reconcile this theory with experiments, it has been necessary to postulate unreasonably large orientation-dependent energies and, by implication, excessively large value of $\Delta\alpha$ [78, 79)]

in the mean field of its neighbors. Here, k is Boltzmann's constant, T^* is the characteristic temperature that measures the strength of the orientation-dependent interaction, and s is the order parameter defined by

$$s = 1 - \frac{3}{2} \langle \sin^2 \psi \rangle \tag{15}$$

Thus, kT^* is the orientation-dependent energy ε^0_{orient} due to interactions between a segment and its neighbors in the perfectly ordered ($\psi = 0$ and $s = 1$) neat fluid. The characteristic temperature is related to $\Delta\alpha$ according to [77]

$$kT^* = Const. \ r_*^{-6}(\Delta\alpha)^2 \tag{16}$$

where r_* is the distance between neighboring segments in contact.

The orientation-dependent energy for the system as a whole, obtained as the averaged sum of the mean field interactions of all segments, is

$$E_{orient} = -\frac{1}{2}xn_p kT^* v_p s^2 \tag{17}$$

as follows from Eq. (14). The factor 1/2 eliminates pairwise redundance.

The direct contribution of the orientation-dependent interactions to the reduced free energy, obtained by division of Eq. (17) by kT, is $-\frac{1}{2}xn_p v_p s^2 T^*/T$. The mean field orientational energy $\varepsilon(\psi)$ also affects the orientational distribution (see Eq. (B-1) of Appendix-B) which, in turn, alters the orientational factor Z_{orient} in the partition function and, hence, makes a further contribution to the reduced free energy [30, 77]. The chemical potentials thus derived are given by Eqs. (B-3) and (B-4) of the Appendix. The equations governing biphasic equilibrium are obtained by equating these expressions to Eq. (A-15) and (A-16), respectively, for the chemical potentials of solvent and polymer in the coexisting isotropic phase.

Calculations for neat fluids are summarized in Fig. 9 where the reciprocal of the reduced nematic-isotropic transition temperature, T_{ni}/xT^*, at equilibrium is plotted against the axial ratio x [77]. The curve represents solutions of the relation obtained by equating Eq. (B-4) for $\mu'_p - \mu^0_p$ in the nematic phase to Eq. (A-16) for $\mu_p - \mu^0_p$ in the isotropic phase, with $v'_p = v_p = 1$. Starting at x = 6.4 for hard rods the calculated curve delineates the orientation-dependent intermolecular attractions required to compensate an axial ratio x < 6.4 in order to establish biphasic equilibrium. This functional relationship has been used to correlate and compare the transition temperatures observed for low molecular nematogens [77]. The axial ratio can be estimated from the molecular structure. The curve in Fig. 9 then gives the corresponding value of xT^*/T_{ni} which, in conjunction with the observed transition temperature T_{ni}, yields the characteristic temperature T^* for the given nematogen. Values of T^* thus obtained are mutually consistent with qualitative inferences from the molecular constitution.

For a more definitive test of the theory, Irvine [79] has determined the optical anisotropies of nematogens and analogous compounds from depolarized Rayleigh

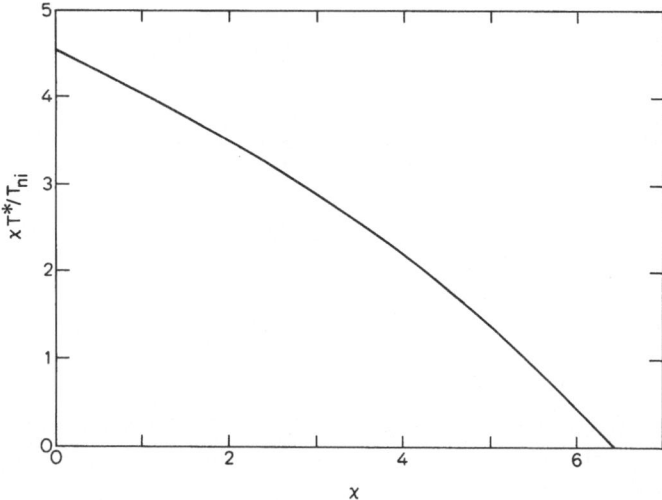

Fig. 9. Reciprocal of the reduced temperature for nematic-isotropic equilibrium plotted against the axial ratio x (From Flory and Ronca [77])

scattering measurements on their dilute solutions. The value of $\Delta\alpha$ determined in this way affords a means for independent estimation of T* through use of the relationship

$$\varepsilon^0_{orient}/\varepsilon_{iso} = \sigma(\Delta\alpha/\bar{\alpha})^2 \qquad (18)$$

giving the ratio of the orientation-dependent energy ε^0_{orient} of interaction of a segment with its neighbors for perfect order (see above) to the corresponding mean intermolecular energy ε_{iso} for a segment in the isotropic fluid as a function of the ratio of $\Delta\alpha$ to the mean polarizability $\bar{\alpha}$. One may calculate $\bar{\alpha}$ from the refractive index of the fluid using the Lorentz-Lorenz relationship. The energy ε_{iso}, being twice the intermolecular energy per segment[9], may be evaluated from the energy of vaporization or from the equation of state [79]. It follows from Eqs. (14) and (18) that

$$kT^* = - \sigma(\Delta\alpha/\bar{\alpha})^2\, \varepsilon_{iso} \qquad (19)$$

From the theory of London dispersion interactions as extended by Inura and Okano [80] and other [78, 81] to include anisotropic interactions, one may estimate $\sigma \approx 0.05$. Values of T* evaluated on this basis according to Eq. (19) are compared in Table 4 with those deduced from the transition temperatures T_{ni}, the axial ratios x, and the function shown in Fig. 9. The agreement is quite satisfactory. The values

---

9 The energies ε as defined here, and in Eq. (14) as well, are twice the energies per segment obtained through division of the corresponding total intermolecular energies by the number of segments. The factor of two enters as the obvious consequence of the fact that the total energy is the sum over interacting pairs, two segments being engaged in each pair. It is to be noted that T* is defined [77] (see Eqs. (14) and (16)) in terms of the energy $\varepsilon(\psi)$ affecting a given segment, without introduction of the factor one-half that would allow for this redundance.

Table 4. [79)] Evaluations of T^* for Low Molecular Nematogens

Nematogen	x	T_{ni} (K)	T^*, K	
			From T_{ni}	Estimated from $\Delta\alpha$
$H(C_6H_4)_5H$	5.0	718	485[b]	550[b]
B-4[a)]	5.2	529	125	150
MBBA	4.2–4.5	319	140	180
PAA	3.6	409	280	300

[a)] $C_6H_5CO(OC_6H_4CO)_2OC_6H_5$
[b)] Calculated with allowance for free volume [79)]

(last column) estimated from the anisotropy $\Delta\alpha$ generally exceed those deduced from T_{ni}, but the differences are well within the bounds of approximations in the theory and of uncertainties in x and other quantities required.

Calculations of biphasic equilibria in lyotropic systems containing a diluent have been carried out using the two equilibrium relations identified above [30)]. Although the orientation-dependent interactions are attenuated by dilution, they may nevertheless affect the character of the phase diagram markedly. For x > ca. 30 and xT^*/T on the order of 0.10, a second (re-entrant) biphasic gap emerges, being preceded by a critical point and followed by a triple point [30)], exactly as found for the analogous system (see Fig. 8) subject to isotropic attractions between the rodlike particles. It is noteworthy that in calculations like those leading to Fig. 8 the isotropic interactions have been assumed to be operative in both phases, as is implicit in the assignment of the same value of χ for the respective phases. They are reduced somewhat in the isotropic phase, relative to the conjugate anisotropic phase, owing to the lower concentration in the former. The anisotropic interactions, on the other hand, necessarily vanish altogether in the isotropic phase. Yet, the same qualitative features appear in both circumstances. According to the calculations, they are more prominent for systems subject to anisotropic interactions, and this may be a consequence of the greater difference between the interactions in the two phases in this case.

The orientation-dependent, "soft" interactions here considered doubtless play an important role in polymers and copolymers exhibiting mesomorphic behavior. Such polymers typically contain aromatic residues, having large anisotropies, as integral members of the chain backbone. These interactions are especially prominent in melts and concentrated solutions, as has been pointed out. Their influences in mesogenic polymers have received little attention.

Appendix

A. Lattice Theory for Hard Rods with Exact Treatment of the Orientation Distribution [7)]

The reduced free energy obtained by combining Eqs. (6) and (8) according to Eq. (7) and replacing factorials by their Stirling's approximations is

$$-\ln Z_M = n_s \ln v_s + n_p \ln (v_p/x) + n_p(\bar{y} - 1)$$

$$- (n_s + \bar{y}n_p) \ln \left[1 - v_p(1 - \bar{y}/x)\right]$$

$$- \sum_y n_{py} \ln (n_p\omega_y/n_{py}) \qquad\qquad\qquad (A\text{-}1)$$

where v_s and v_p are volume fractions of solvent and rodlike solute, respectively. For perfect order, $\bar{y} = 1$ and Eq. (A-1) reduces to

$$-\ln Z_M = n_s \ln [n_s/(n_s + n_p)] + n_p \ln [n_p/(n_s + n_p)] - n_p \ln \omega_1$$

which is the expression for ideal mixing, apart from the additive term $-n_p \ln \omega_1$. Reduction to the ideal mixing law is a consequence of the one-dimensionality of mixing in this limit. For random disorder, $n_{py}/n_p = \omega_y$ and $\bar{y} = x$ (see below); hence

$$-\ln Z_M = n_s \ln v_s + n_p \ln (v_p/x) + n_p(x - 1)$$

which is the expression for an athermal polymer solution apart from absence of the term from the disorientation entropy, deliberately ignored above.

According to Eq. (4) of the text, the expected number of situations for molecule j depends exponentially on y_j. The corresponding expression for the full complement of n_p rodlike molecules is obtained by replacing j by n_p in that Equation. Then the expected number of situations for a molecule with disorientation y takes the form

$$v_y = \text{Const. exp} (-ay) \qquad\qquad\qquad (A\text{-}2)$$

where y_j and v_j have been replaced by y and v_y, respectively, and

$$a = -\ln \left[1 - (x - \bar{y}) n_p/n_0\right]$$

$$= -\ln \left[1 - v_p(1 - \bar{y}/x)\right] \qquad\qquad\qquad (A\text{-}3)$$

as follows from Eq. (4). At equilibrium, the relative number of rods with disorientation y is proportional to the product of ω_y and v_y. Hence, the distribution of orientations at equilibrium is given by

$$n_{py}/n_p = f_1^{-1}\omega_y \exp (-ay) \qquad\qquad\qquad (A\text{-}4)$$

where f_1 is the normalization factor.

To proceed it is necessary to relate y to the polar angle ψ with respect to the preferred axis. The diagrams shown for two dimensions in Fig. 1 suggest that $y = x \sin \psi$. In three dimensions, however, the number y of sequences depends on the azimuthal orientation φ of the rod about the preferred axis [7]. For specified values of ψ and φ the value of y is given by the sum of the projections of x on axes perpendicular to the domain axis, i.e.,

$$y_{\psi, \varphi} = x \sin \psi(|\cos \varphi| + |\sin \varphi|) .$$

Upon averaging over φ, one obtains [7]

$$y = (4x/\pi) \sin \psi \tag{A-5}$$

Hence

$$\bar{y} = (4x/\pi) \langle \sin \psi \rangle \tag{A-6}$$

For complete disorientation $\langle \sin \psi \rangle = \pi/4$ and $\bar{y} = x$, as is required.

Inasmuch as ω_y is a measure of the solid angle associated with y, it follows that

$$\omega_y = \sin \psi (d\psi/dy) \tag{A-7}$$

Substitution of Eqs. (A-5) and (A-7) in Eq. (A-4) gives

$$n_{p\psi} \, d\psi = n_p f_1^{-1} \sin \psi \exp(-\alpha \sin \psi) \, d\psi \tag{A-8}$$

where

$$\alpha = (4/\pi) \, ax \tag{A-9}$$

and

$$f_1 = \int_0^{\pi/2} \sin \psi \exp(-\alpha \sin \psi) \, d\psi \tag{A-10}$$

Since

$$\langle \sin \psi \rangle = f_2/f_1$$

where

$$f_2 = \int_0^{\pi/2} \sin^2 \psi \exp(-\alpha \sin \psi) \, d\psi \tag{A-11}$$

it follows from Eq. (A-6) that

$$\bar{y} = (4x/\pi) \, f_2/f_1 \tag{A-12}$$

The reduced chemical potentials obtained by partial differentiation of Eq. (A-1) are [7], [10]

$$(\mu_s - \mu_s^0)/RT = \ln(1 - v_p) + v_p(\bar{y} - 1)/x$$
$$- \ln[1 - v_p(1 - \bar{y}/x)]$$
$$= \ln(1 - v_p) + v_p(\bar{y} - 1)/x + a \tag{A-13}$$

---

10 In performing the partial differentiations it is expedient to consider the ratios n_{py}/n_p in Eq. (A-1) to have their equilibrium values. Then $- \ln Z_M$ is stationary with respect to these ratios, hence derivatives with respect to them are null and may be dismissed. It is essential to recognize, however, that the chemical potentials thus obtained apply only at orientational equilibrium

and

$$(\mu_p - \mu_p^0)/RT = \ln (v_p/x) + v_p(\bar{y} - 1) - \ln f_1 \tag{A-14}$$

the latter equation having been simplified through substitution of Eq. (A-4) for n_{py}/n_p in the argument of the logarithm obtained by differentiation of Eq. (A-1).

Corresponding reduced potentials for an isotropic phase are the familiar expressions from polymer solution theory:

$$(\mu_s - \mu_s^0)/RT = \ln (1 - v_p) + (1 - 1/x) v_p \tag{A-15}$$

$$(\mu_p - \mu_p^0)/RT = \ln (v_p/x) + (x - 1) v_p \tag{A-16}$$

Substitutions of Eqs. (A-13) and (A-14) for μ_s' and μ_p' and of Eqs. (A-15) and (A-16) for μ_s and μ_p in Eq. (12) yield the equations that determine the compositions of the phases coexisting at equilibrium. They are obviously not explicitly soluble. The necessity to evaluate the integrals f_1 and f_2 by numerical integration further complicates their solution. Although numerical results are readily obtainable, the equations do not lend themselves to straightforward manipulations.

B. Rods with Orientation-Dependent Interactions [77]

The interaction energy in the mean field approximation is given by Eq. (14) of the text. The orientation distribution function, Eq. (A-8), is then replaced by [30, 77]

$$n_{p\psi} \, d\psi = n_p f_1^{-1} \sin \psi \exp [- \alpha \sin \psi - (3xv_p s/2\Theta) \sin^2 \psi] \, d\psi \tag{B-1}$$

where $\theta = T/T^*$ and f_1 is redefined by

$$f_1 = \int_0^{\pi/2} \sin \psi \exp [-\alpha \sin \psi - (3xv_p s/2\Theta) \sin^2 \psi] \, d\psi \tag{B-2}$$

with f_2 redefined analogously. The reduced free energy is obtained from Eq. (A-1) by appending the term $v_p x n_p s^2/2\theta$ (see Eq. (17). The chemical potentials are [30]

$$(\mu_s - \mu_s^0)/RT = \ln (1 - v_p) + v_p(\bar{y} - 1)/x + a + v_p^2 s^2/2\Theta \tag{B-3}$$

and

$$(\mu_p - \mu_p^0)/RT = \ln (v_p/x) + v_p(\bar{y} - 1) - \ln f_1 - (xv_p s/\Theta) (1 - v_p s/2) \tag{B-4}$$

Those for the isotropic state in which $s = 0$ are, of course, unaffected; they are given by Eqs. (A-15) and (A-16).

C. The 1956 Approximate Treatment [6)]

Approximation of the orientation factor Z_{orient} in the partition function according to Eq. (9) of the text provides the key to the earlier version of the lattice theory of rodlike particles not subject to orientation-dependent interactions. Although approximate, this formulation offers advantages of simplicity that for most purposes outweigh the incident errors. The latter are generally small (see Table 1).

Combination of Eqs. (6) and (9) according to Eq. (7) yields for the reduced free energy of the mixture

$$-\ln Z_M = n_s \ln v_s + n_p \ln (v_p/x) + n_p(\bar{y} - 1)$$

$$- (n_s + \bar{y}n_p) \ln [1 - v_p(1 - \bar{y}/x)]$$

$$- 2n_p \ln (\bar{y}/x) + \chi x n_p v_s \tag{C-1}$$

where the final term has been included to respresent exchange interactions (isotropic) between the components. For systems of rods devoid of "soft" interactions, $\chi = 0$. The contribution from the disorientation entropy is not included in Eq. (C-1), it having been absorbed into the state of reference (see Eq. (9) and footnote 3).

Minimization of $-\ln Z_M$ with respect to $\bar{y}$ at fixed composition yields

$$2/y = -\ln [1 - v_p(1 - \bar{y}/x)] = a \tag{C-2}$$

a having been defined in Eq. (A-3). Equation (C-2) is equivalent to Eq. (10). This equation supplants the equilibrium orientation distribution function, Eq. (A-4), and thus circumvents the integrals f_1 and f_2.

Partial differentiation of Eq. (C-1) yields the reduced chemical potentials

$$(\mu_s - \mu_s^0)/RT = \ln (1 - v_p) + v_p(\bar{y} - 1)/x + a + \chi v_p^2 \tag{C-3}$$

$$= \ln (1 - v_p) + v_p(\bar{y} - 1)/x + 2/y + \chi v_p^2 \tag{C-3'}$$

and

$$(\mu_p - \mu_p^0)/RT = \ln (v_p/x) + v_p(\bar{y} - 1) + 2[1 - \ln (\bar{y}/x)] + \chi x(1 - v_p)^2 \tag{C-4}$$

Equation (C-3) is identical to Eq. (A-13) apart from the exchange interaction term. The corresponding chemical potentials under conditions of isotropy are

$$(\mu_s - \mu_s^0)/RT = \ln (1 - v_p) + (1 - 1/x) v_p + \chi v_p^2 \tag{C-5}$$

and

$$(\mu_p - \mu_p^0)/RT = \ln (v_p/x) + (x - 1) v_p + \chi x(1 - v_p)^2 \tag{C-6}$$

which reduce to Eqs. (A-15) and (A-16) for $\chi = 0$.

The 1956 approximation is generally inapplicable to systems in which orientation-dependent interactions must be taken into account [77].

Acknowledgment: This work was supported by the Directorate of Chemical Sciences, U.S. Air Force Office of Scientific Research Grant No. AFOSR 82-0009.

References

1. S. Chandrasekhar: "Liquid Crystals", Cambridge Univ. Press (1977)
2. P. G. de Gennes: "The Physics of Liquid Crystals", Clarendon Press, Oxford (1971)
3. "Liquid Crystalline Order in Polymers", Ed. by A. Blumstein: Academic Press, New York (1978)
4. "Polymer Liquid Crystals", A. Ciferri, W. R. Krigbaum and R. B. Meyers, Eds.: Academic Press, New York (1982)
5. W. Maier and A. Saupe: Z. Naturforschg. *14a*, 882 (1959); ibid., *15a*, 287 (1960)
6. P. J. Flory: Proc. Royal Soc., London *A 234*, 73 (1956)
7. P. J. Flory and G. Ronca: Mol. Cryst. Liq. Cryst. *54*, 289 (1979)
8. L. Onsager: Ann. N.Y. Acad. Sci. *51*, 627 (1949)
9. P. J. Flory and A. Abe: Macromolecules *11*, 1119 (1978)
10. C. Robinson: Trans. Faraday Soc. *52*, 571 (1956). C. Robinson, J. C. Ward and R. B. Beevers: Discuss. Faraday Soc. *25*, 29 (1958)
11. J. Hermans, Jr.: J. Coll. Sci. *17*, 638 (1962)
12. A. Nakajima, T. Hayashi and M. Ohmori: Biopolymers *6*, 973 (1968)
13. E. L. Wee and W. G. Miller: "Liquid Crystals and Ordered Fluids", Vol. 3. Ed by J. F. Johnson and R. S. Porter, Plenum Press, 1978, p. 371. W. G. Miller, C. C. Wu, E. L. Wee, G. L. Santee, J. H. Rai and K. D. Goebel: Pure Appl. Chem. *38*, 37 (1974)
14. G. Kiss and R. Porter: J. Polym. Sci. Polym. Symp. *65*, 193 (1978)
15. S. M. Aharoni and E. K. Walsh: Macromolecules *12*, 271 (1979); J. Polym. Sci. Polym. Lett. Ed. *17*, 321 (1979)
16. R. R. Matheson, Jr.: Macromolecules *13*, 643 (1980)
17. S. M. Aharoni: Polymer *21*, 1413 (1980)
18. S. P. Papkov, V. G. Kulichikhin, V. D. Kalmykhova and A. Ya. Malkin: J. Polymer Sci., Polymer Physics Ed. *12*, 1753 (1974). V. G. Kaluchikhin, S. L. Papkov et al.: Polym. Sci. USSR *18*, 672 (1976)
19. S. P. Papkov: Polym. Sci. USSR *19*, 1 (1977)
20. J. R. Schaefgen et al.: Polym. Prepr. Am. Chem. Soc. Div. Polym. Chem. *17* (1), 69 (1976)
21. S. L. Kwolek, P. W. Morgan, J. R. Schaefgen and L. W. Gulrich: Macromolecules *10*, 1390 (1977)
22. C. Balbi, E. Bianchi, A. Ciferri and A. Tealdi: J. Polym. Sci., Polym. Phys. Ed. *18*, 2037 (1980)
23. T. I. Bair, P. W. Morgan and F. L. Kilian: Macromolecules *10*, 1396 (1977)
24. A. Nakajima, T. Hirai and T. Hayashi: Polym. Bull. *1*, 143 (1978)
25. S. M. Aharoni: Macromolecules *12*, 94 (1979)
26. S. M. Aharoni: J. Polym. Sci. Polym. Phys. Ed. *18*, 1439 (1980)
27. S. M. Aharoni: Polym. Bull. *9*, 186 (1983)
28. A. J. Bur and L. J. Fetters: Chem. Rev. *76*, 727 (1976)
29. A. A. Brian, H. L. Frisch and L. S. Lerman: Biopolymers *20*, 1305 (1981)
30. M. Warner and P. J. Flory: J. Chem. Phys. *73*, 6327 (1980)
31. R. R. Matheson, Jr., and P. J. Flory: Macromolecules *14*, 954 (1981)
32. A. Abe and P. J. Flory: Macromolecules *11*, 1122 (1978)
33. P. J. Flory and R. S. Frost: Macromolecules *11*, 1126 (1978)
34. R. S. Frost and P. J. Flory: Macromolecules *11*, 1134 (1978)
35. J. K. Moscicki and G. Williams: Polymer 22, 1451 (1981); ibid. *23*, 558 (1982)
36. G. Conio, E. Bianchi, A. Ciferri and A. Tealdi: Macromolecules *14*, 1084 (1981)
37. M. Ballauff, D. C. Wu and P. J. Flory: in preparation

38. M. Ballauff and P. J. Flory: in preparation
39. P. J. Flory: Macromolecules *11*, 1138 (1978)
40. S. M. Aharoni: Polymer *21*, 21 (1980)
41. E. Bianchi, A. Ciferri and A. Tealdi: Macromolecules *15*, 1268 (1982)
42. W. F. Hwang, D. R. Wiff and C. L. Benner: J. Macromol. Sci. Phys. *B 22*, 231 (1983)
43. D. A. Brant and K. D. Goebel: Macromolecules *5*, 536 (1972). K. D. Goebel, C. E. Harvie and D. A. Brant: Applied Polymer Symp. *28*, 671 (1976)
44. A. Ciferri: Chapter 3 of ref. 4
45. A. Roviello and A. Sirigu: Makromol. Chem. *183*, 895 (1982); see also references cited therein
46. P. J. Flory: Prof. Royal Soc., London *A 234*, 60 (1956)
47. G. Ronca and D. Y. Yoon: J. Chem. Phys. *76*, 3295 (1982)
48. P. J. Flory: Macromolecules *11*, 1141 (1978)
49. S. M. Aharoni: Mol. Cryst. Liq. Cryst. Lett. *56*, 237 (1980)
50. S. Dayan, P. Maissa, M. J. Vellutini and P. Sixou: J. Polym. Sci. Polym. Lett. Ed. *20*, 33 (1982)
51. J. Bheda, J. F. Fellers and J. L. White: Coll. Polym. Sci. *258*, 1335 (1980)
52. R. S. Werbowyj and D. G. Gray: Macromolecules *13*, 69 (1980)
53. G. Conio, E. Bianchi, A. Ciferri, A. Tealdi and M. A. Aden: Macromolacules *16*, 1264 (1983)
54. H. Chanzy, A. Peguy, S. Chaunis and P. Monzie: J. Polym. Sci. Polym. Phys. Ed. *18*, 1137 (1980)
55. D. L. Patel and R. D. Gilbert: J. Polym. Sci. Polym. Phys. Ed. *19*, 1231 (1981)
56. D. W. Tanner and G. C. Berry: Polym. Sci. Polym. Phys. Ed. *12*, 941 (1974)
57. K. Kamide, Y. Miyazaki and T. Abe: Polym. J. *11*, 523 (1979)
58. P. Navard, J. M. Haudin, S. Dayan and P. Sixou: J. Polym. Sci. Polym. Lett. Ed. *19*, 379 (1981)
59. P. J. Flory, O. K. Spurr, Jr., and D. K. Carpenter: J. Polym. Sci. *27*, 231 (1958)
60. M. G. Wirick and M. H. Waldman: J. Appl. Polym. Sci. *14*, 579 (1970)
61. G. Maret, M. Milas and M. Rinaudo: Polymer Bull. *4*, 291 (1981). M. Rinaudo and M. Milas: Carbohydrate Polymers *2*, 264 (1982). M. Milas and M. Rinando: Polym. Bull. *10*, 271 (1983)
62. K. Van, T. Norisuye and A. Teramoto: Mol. Cryst. Liq. Cryst. *78*, 123 (1981)
63. G. M. Holtzer, Chap. 2 in "Solution Properties of Polysaccharides", D. A. Brant, Ed.: Am. Chem. Society Sympos. Series, No. 150
64. E. D. T. Atkins and K. D. Parker: J. Polym. Sci., Part C *28*, 69 (1969). Proc. Royal Soc. London, Sect. B *173*, 209 (1969)
65. T. Norisuye, T. Yanaki and H. Fujita: J. Polym. Sci. Polym. Phys. Ed. *18*, 547 (1980)
66. T. Yanaki, T. Norisuye and H. Fujita: Macromolecules *13*, 1462 (1981)
67. Y. Kashiwagi, T. Norisuye and H. Fujita: Macromolecules *14*, 1220 (1981)
68. J. I. Jin, S. Antoun, C. Ober and R. W. Lenz: Brit. Polym. J. *12*, 132 (1980). C. Ober, J. I. Jin, R. W. Lenz: Polymer J. *14*, 9 (1982)
69. A. Abe: in preparation
70. W. J. Jackson and H. F. Kuhfuss: J. Polym. Sci., Polym. Chem. Ed. *14*, 2043 (1976)
71. C. Viney, G. R. Mitchell and A. H. Windle: Polym. Commun. *24*, 146 (1983). A. M. Donald, C. Viney and A. H. Windle, Polymer *24*, 155 (1983)
72. G. R. Mitchell and A. H. Windle: Polymer *23*, 1269 (1982)
73. R. R. Matheson, Jr., and P. J. Flory: in preparation
74. R. A. Orwoll and P. J. Flory: in preparation
75. S. Ya. Frenkel, L. G. Shaltyko and G. K. Elyashevich: J. Polym. Sci., Part C *30*, 47 (1970)
76. P. Pincus and P. G. de Gennes: J. Polym. Sci. Polym. Symp. *65*, 85 (1978)
77. P. J. Flory and G. Ronca: Mol. Cryst. Liq. Cryst. *54*, 311 (1979)
78. A. Wulf: J. Chem. Phys. *64*, 104 (1976)
79. P. A. Irvine and P. J. Flory: in press.
80. H. Imura and K. Okano: J. Chem. Phys. *58*, 2763 (1973)
81. M. Warner: J. Chem. Phys. *73*, 5874 (1980)

M. Gordon (editor)
Received June 12, 1983

Polypeptide Liquid Crystals

Ichitaro Uematsu
Tokyo Institute of Technology, Meguro, Tokyo, 152, Japan
Yoshiko Uematsu
Tokyo Institute of Polytechnics, Atsugi, Kanagawa, 243-02, Japan

The recent studies on the structure and properties of polypeptide liquid crystals, which are formed in solution as well as in the solid state, are reviewed in this article. Especially the cholesteric pitch and the cholesteric sense (right-handed or left-handed), which are characteristic factors of cholesteric liquid crystals, are discussed in detail in relation to the effects of temperature, concentration, and solvent. Further cholesteric liquid crystalline structure retained in cast films and thermotropic mesomorphic state in some copolypeptides are also discussed.

1 Formation of Liquid Crystals in Polypeptide Solutions 38
  1.1 Introduction 38
  1.2 Pretransitional Regions 40
  1.3 Biphasic Regions 41
      1.3.1 Narrow Biphasic Regions 41
      1.3.2 Wide Biphasic Regions 41
  1.4 Effect of External Fields on the Phase Transition 45

2 Cholesteric Liquid Crystals in Polypeptide Solutions 46
  2.1 Optical Properties of Cholesteric Liquid Crystals 46
  2.2 Dependence of the Cholesteric Pitch on Temperature and Concentration 52
  2.3 Solvent Effects 58

3 Cholesteric Twisted Structure in Solid Films of Polypeptides 66

4 Concluding Remarks 71

5 References 71

Advances in Polymer Science 59
© Springer-Verlag Berlin Heidelberg 1984

1 Formation of Liquid Crystals in Polypeptide Solutions

1.1 Introduction

Liquid crystals of polypeptide solutions were found by Elliott and Ambrose [1] in 1950. They observed a birefringent solution phase when preparing films of poly(γ-benzyl L-glutamate) (PBLG) from chloroform solutions. Robinson [2] investigated the birefringent solution extensively and found that it showed properties quite similar to those of low molar mass cholesteric liquid crystals. Furthermore he concluded that α-helical polypeptide molecules aligned nearly parallel in a layer and that the layer stacks successively changing the alignment direction by a constant angle around the normal of the layer.

The necessary condition of polymer solutions to form liquid crystals is firstly that polymers are rigid rod-like molecules. This is clear from the fact that RNA, tobacco mosaic virus and aromatic polyamides manifest liquid crystalline structure. Furthermore, to realize the liquid crystalline state, polymer solutions must be concentrated above a critical concentration. PBLG solutions above the critical concentration separate into two phases: a birefringent phase with high polymer concentration separates in an isotropic phase with low concentration. At a higher concentration the solution becomes a uniform birefringent phase. Robinson denoted the incipient concentration for the appearance of the birefringent phase and the concentration for the completion of the uniform birefringent phase as A and B points, respectively. A and B points are dependent on the molecular mass of the polymer and are not much influenced by solvent properties. These results suggest that the formation of liquid crystals in polymer solutions is related to the rod-like structure of polymers and that the molecular interaction between polymer and solvent is not so important. The phase separation or self-ordering in the solution of rod-like molecules has been suggested by Onsager [3] and Flory [4] and since then many theoretical treatment has been published [5]. Flory applied a lattice model treatment to the system consisting of impenetrable rod molecules without molecular interaction and sphere-like solvent molecules. He showed theoretically that the rod molecules which arrange randomly in dilute solutions, tend to align parallel with the increase of the concentration and that the liquid-liquid phase separation takes place at a critical concentration.

According to the Flory theory, activities a_1, a_2 are given in the following equations.

Isotropic phase

$$\ln a_1 = \ln (1 - v_2) + [(r - 1)/r]v_2 + xv_2^2 \tag{1}$$

$$\ln a_2 = \ln v_2 + (r - 1)v_2 - \ln r^2 + xr(1 - v_2)^2 \tag{2}$$

Anisotropic phase

$$\ln a_1 = \ln (1 - v_2^*) + [(y - 1)/r]v_2^* + 2/y + xv_2^{*2} \tag{3}$$

$$\ln a_2 = \ln v_2^* + (y - 1)v_2^* + 2 - \ln y^2 + xr(1 - v_2^*)^2 \tag{4}$$

where r and x are the axial ratio of the rod-like polymer and the polymer solvent interaction parameter, respectively. y is a orientation parameter given by the following equation and is unity at higher concentration.

$$v_2^* = [r/(r - y)] [1 - exp(-2/y)] \qquad (5)$$

v_2, v_2^* is the equilibrium concentrations (volume fraction) of the isotropic and anisotropic phase respectively.

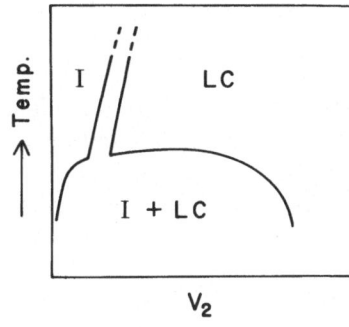

Fig. 1. Temperature-composition phase diagram of polypeptide solutions

The phase diagram of the polypeptide solutions is shown schematically in Fig. 1. According to the Flory theory, the relation between the concentration at the A point (v_{2A}) and the axial ratio (r) is represented as follows;

$$v_{2A} = (8/r)[1 - (2/r)] \qquad (6)$$

A and B points are easily determined by polarizing microscopic observation of polymer solutions. The Flory theory is in fairly good agreement with experimental results despite the approximations included in theoretical treatments. This shows that the origin of the formation of liquid crystals in rigid rod-like polymer solutions is entirely geometrical. Recently Flory et al. [7] extended the theory to polydisperse mixtures and phase equilibria have been calculated for various model systems. Partitioning of the solute components between anisotropic and isotropic phase and the biphasic gap have been discussed. Furthermore they presented a theory for a system of rigid, impenetrable, rod-like molecules subject to orientation-dependent mutual attraction.

However, only limited experimental studies on the thermodynamic properties of polypeptide solutions have been carried out. The results of vapor sorption studies for PBLG and poly(β-benzyl L-aspartate) solutions at high polymer concentrations by Flory and Leonard [8] could not be explained by the Flory model, but could be explained by assuming that mixing of solvent with flexible side chains dominates the thermodynamic behavior at high concentrations. Rai and Miller [9] obtained similar results for the PBLG-dimethylformamide (DMF) system at high concentrations. They also showed that the results could be explained by the Wee-Miller [10] theory in which modification of Flory's lattice theory to allow for side chain flexibility had been made. The statistical thermodynamics was viewed in two points; one involving the configurations of the rods on the lattice with no rod-solvent interaction, the other involving the configurational statistics of placing flexible side chains on the lattice, and their interaction with the solvent.

Kubo and Ogino [11] have measured solvent activities in the PBLG-chloroform system. The data could be explained by the Flory theory modified by Wee and Miller, in a low or intermediate concentration range, and by the Flory-Leonard model at high concentrations. Further, they showed that the difference in the concentration dependence of osmotic pressures between the isotropic and anisotropic phases was not so large as predicted by theories.

1.2 Pretransitional Regions

When the polypeptide concentration exceeds the critical concentration at the A point, the solution separates into two phases. The polypeptide-rich phase is birefringent and separates from the more dilute medium initially in the form of spherical liquid droplets or spherulites. If the two-phase solution is cooled or further concentrated in polymer, the spherulites grow in size and coalesce forming a continuously birefringent fluid at the B point.

Patel and DuPré [12] have investigated the optical dispersion of dilute solutions of PBLG and the enantiomer PBDG in various solvents with the concentration of polymer held just below the critical volume fraction for the appearance in the polarizing microscope of cholesteric spherulites. In this concentration regime, solutions are uniformly birefringent and strongly optically active. Some light passes through the sample when viewed between crossed polars and a uniform grainy texture is observed, without however any evidence of phase separation. They found the nature of the optical rotatory dispersion (ORD) behavior in the pretransition region to be strongly dependent on the nature of the solvent. The effects observed fall into three classes:

1) Both PBLG and PBDG solution have the same chiral sense in the pretransition region. The signs of the optical rotations are predominantly negative. As the temperature is increased the solution behavior approaches that of the isotropic liquid. This effect is observed only in chloroform.

2) The chirality of the pretransition liquid crystal is the same as that of the dilute solution of the respective polymer. As the temperature is increased, the optical activity is reduced in a continuous fashion to that observed in dilute solution, without change of sign. This effect is observed in CH_2Cl_2, 1,2-dichloroethane (EDC), pyridine, and 2-chloropyridine.

3) The chirality of the pretransition phase is opposite to that of the dilute solution. As the temperature increases, ORD curves reverse sense from predominantly positive to negative for PBLG solutions, and from predominantly negative to positive for PBDG solutions. This effect is observed in dioxane, DMF, tetrahydrofuran, 1,1,2,2-tetrachloroethane, and 1,3-dioxolane.

An analogous enhancement in the optical rotatory power of the isotropic phase of a thermotropic liquid crystal has been observed near the isotropic-cholesteric phase transition. Patel and DuPré concluded that it is attributable to short-range chiral ordering of the long axes of the macromolecules.

1.3 Biphasic Regions

1.3.1 Narrow Biphasic Regions

A small change in polymer concentration will cause the solution in the pretransition region to turn completely isotropic or will precipitate the cholesteric spherulites. A small change in temperature will also cause the same effects. As the A point is passed, spherulites become visible and grow in size. The series of light and dark rings observed within the spherulites are equidistant and correspond to the arrangement of cholesteric planes each warped into a spherical surface. Optical retardation measurements indicate that the macromolecules are tangential to these "onion-shell" layers. Patel and DuPré [13] have found that rings are notably absent in spherulites formed from solutions of higher molecular-weight polymer. They attributed it to the cholesteric pitch of the liquid crystal being larger than or of the same order as the radius of the droplet. In these cases the internal organization of the macromolecules will be governed by the surface energy of the droplet and the cumulative twist will not be developed until the droplets coalesce.

The size of the spherulites depends on the molecular mass of the polymer and the density of the solvent. Higher molecular-mass polypeptides generally yield smaller spherulites which do take a longer time to develop. If the density of the solvent and polymer are nearly the same, the spherulites tend to grow to larger dimensions. If the density of the solvent is less than that of the polymer, smaller spherulites are formed. In this case, the spherulite will collide with the cell wall and collapse before its growth is complete.

Miller et al. [14] have initiated studies to elucidate the kinetics of forming the ordered phase in the polypeptide solutions. When the isotropic solution is temperature-jumped across the biphasic region into the region in which the ordered phase is stable, the kinetics can be described by a nucleation and growth mechanism with many similarities to the kinetics of polymer crystallization. They have also shown that the kinetic process can be divided into two time scales: the conversion of the randomly oriented rods to a random array of locally oriented rod domains, followed by growth of some domains at the expense of others.

When the temperature is adjusted so that the final state is in the narrow biphasic region of the phase diagram, ordered spherulites appear in the disordered phase. Although the time scale becomes long as the driving force to form the ordered phase becomes small, the appearance of spherulites suggests a nucleation mechanism.

1.3.2 Wide Biphasic Regions

When an isotropic, biphasic, or ordered solution is brought into the wide biphasic region of the phase diagram, the solutions are always observed to form a transparent, mechanically self-supporting gel. The gel formation is concentration and temperature dependent and completely reversible. Gelation occurs in the PBLG-DMF system $10 \sim 15°$ below the temperature at which light is first observed coming through crossed polars, which Miller et al. have designated previously as the phase boundary separating the isotropic from the biphasic region.

A mechanically self-supporting gel implies a three dimensional network structure. In our system there is no possibility of covalent crosslinks, of noncovalent crosslinks

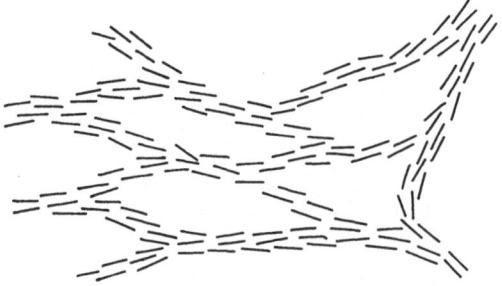

Fig. 2. Model for networks formed by aggregation of rigid rodlike polymers

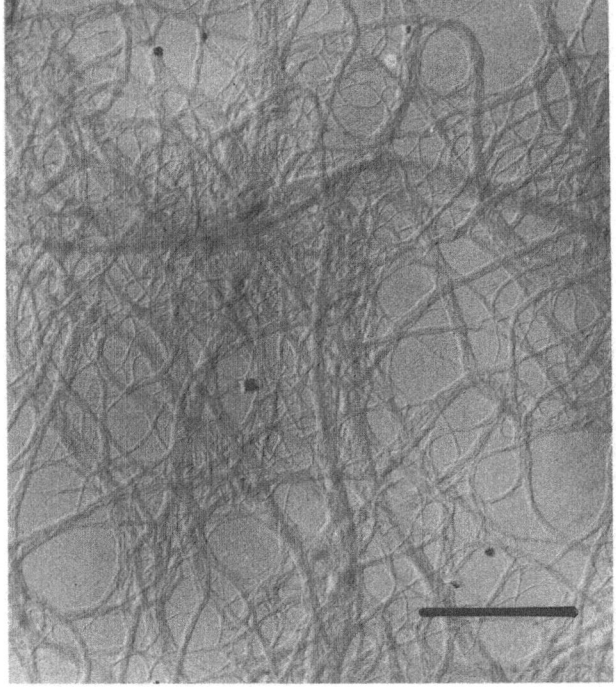

Fig. 3. Electron micrograph of a PBLG-benzylalcohol gel with c = 0.05 wt% (bar = 1 μm)

such as hydrogen bonds, or of entrapment of polymer chains between crystallites to give pseudocrosslinks. However, since the tendency to parallel, side-by-side aggregation of PBLG is considered to be strong near the gelation temperature, the formation of long strands containing many polypeptide molecules is easily attained. For an infinite network to be formed, these strands must be crosslinked. The formation of the cross-links takes place by the branching and rejoining of different sheaflike aggregates throughout the solution (Fig. 2). Fig. 3 shows a transmission electron microscopic photograph of the aggregate formed in the 0.05 wt% gel. The fibrils in the photograph appear to be aggregates of microfibrils 100 Å or less in diameter.

Luzzati et al. [15] suggested the presence of coiled coils in the so-called "complex" phase, described by them and found in the system PBLG-DMF from about 15 to 70 (wt/wt) PBLG, where the position of the reflections in the X-ray diffraction measurement is found to be independent of concentration; in other words, the phase contains a fixed amount of solvent. Parry and Elliott [16] in their more detailed examination of this phase have concluded that coiled coils are not present and that the presence of a strong near-equatorial streak, which is one of the main features expected in the diffraction pattern from a coiled coil, does not necessarily demand the presence of a coiled coil structure. They also found that benzyl groups stack to produce a quasi-helical arangement with different periodicity from that of the α-helix of the main chain and that stacks on neighboring molecules link up to form a continuous quasi-helical side chain configuration. It is possible that a molecular association caused by the benzenring interaction or stack proposed by Parry and Elliott is responsible for the gelation.

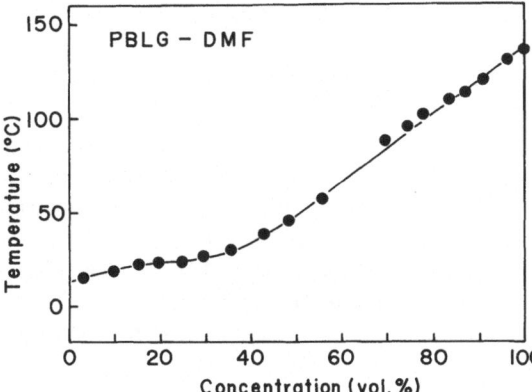

Fig. 4. Melting point-concentration relation for the PBLG-DMF system

In Fig. 4 transition temperatures at peaks in the DSC thermograms are plotted against concentration for the PBLG-DMF system [17]. Sharp X-ray reflections appear in a concentration range from 20 to 60 wt%, the spacing being constant and independent of the total concentration. Therefore, the gel has a crystalline character. The apparent shape of the unit cell in the two-dimensional projection is hexagonal with a side a = 35 Å. An analysis of the complex structure has been attempted on models in which three polymer chains passed through a unit cell [18]. However, it was found recently that four chains pass through the cell [17]. The fact that four chains form a structural unit suggests the possibility of a fundamental microfibril. X-ray diffraction patterns of gels with concentrations less than 20 wt% are smeared out by the scattering from the solvent. However, the continuity of the melting point curve suggests that the complex phase is formed even in dilute gels.

The gel with a concentration higher than 60 wt% showed a broad transition and the transition was irreversible. The solid film has the structure of form A reported by McKinnon and Tobolsky [19]. They have reported that films cast from DMF have two solid state modifications (form A and B) depending on the casting temperature, which are quite different from films cast from chloroform (form C). The characteristic aspects of these modifications in viscoelastic properties are: higher modulus in the temperature range from 30 to 140 °C and an abrupt decrease in modulus near 120 °C for form A, an abrupt decrease of modulus near 90 °C for form B, and a decrease of modulus at room temperature and no considerable decrease of modulus until 110 °C for form C. We also have investigated the effect of solvent on the structure and properties of cast films of PBLG [20]. Cast films are classified into two types: film A cast from the EDC (1,2-dichloroethane) series (e.g., EDC, CH_2Cl_2, and $CHCl_3$) and film B cast from the benzene series (e.g., benzene and tetrahydrofuran). As for viscoelastic properties, film A shows peaks of tan δ at 40 and 105 °C (110 Hz). In the case of film B, the lower temperature peak attributable to the onset of side chain motions appears at a higher temperature and the peak is broadened. Film B further exhibits a decrease of modulus at about 135 °C, which is accompanied by a small peak of tan δ at this temperature. The results of film B are similar to the behavior of form A reported by McKinnon and Tobolsky. The DSC thermogram for film B (form A) shows an endothermic peak at 135 °C, which for film A such a transition is not observed. After being annealed at 140 °C, film B does not show the endothermic peak.

The aggregates of "superhelices" of PBLG, which are precipitated from DMF solution by adding propionic acid, also show an endothermic peak at 135 °C [21,22]. In Fig. 5 a typical electron micrograph of the superhelical aggregates is shown. Ishikawa and Kurita [23] and Tachibana and Kambara [24] have investigated the morphology of precipitates from DMF-propionic acid solutions by means of electron micrography and X-ray diffraction analysis. Ishikawa and Kurita called the rope-like fibrils which are observed by the electron micrography, "super-helices". Tachibana and Kambara reported that the poly(γ-benzyl glutamate) in the precipitates had the conformation of an α-helix and the superhelices had the sense opposite to that of the

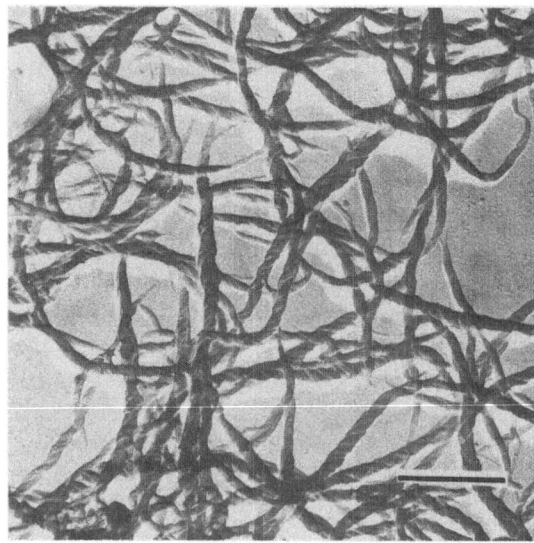

Fig. 5. Electron micrograph of super-helical aggregate of PBLG (bar = 1 μm)

α-helix. The X-ray diagram of the superhelical aggregates is very similar to that of film B. From this results as well as from experimental results of thermal measurements, we can presume that in films cast from solvents of benzene series including DMF, a superhelical structure is formed. Furthermore we can presume that the molecular association in the solution caused by the benzene-ring interaction or stack proposed by Parry and Elliot is responsible for the formation of superhelices and that the film cast from solvents of the benzene series retains the structure existing in the solution. The irreversible transition observed in the thermal measurement of form A is ascribed to the dissociation of the benzene-ring stack in the side chain.

Melting phenomena of PBLG-DMF gel shown in Fig. 4 are classified into three categories depending on the polymer concentration [25]. The melting of dilute gels (C < 18 wt%) is a transition to the isotropic solution. The transition of concentrated gels (20 ~ 60 wt%) is the mixing of the complex phase and the solvent-rich phase into the liquid crystal. The complex phase involves the stacking of side-chain benzene rings. The transition of more concentrated gels (C > 70 wt%) is accompanied by the collapse of this stacking structure.

1.4 Effect of External Fields on the Phase Transition

Recent developments in ultrahigh-modulus polymers have stimulated considerable interest in possible methods for obtaining materials composed of fully stretched, fully oriented macromolecules. An approach for achieving orientation of rigid macromolecules is based on the occurrence of nematic solutions or melts where anisotropic properties are exhibited even in the absence of a flow field. Marrucci et al. [26] investigated the effect of elongational flow on the isotropic-anisotropic transition. This has been done by extending Flory's theory of phase equilibria of rod-like molecules to the case in which the equilibrium is modified by the application of an extensional flow field. They used a quasi-equilibrium statistical mechanical approach, assuming that the velocity field was described in terms of a potential. They showed how the boundaries of the two phase region change as a function of elongational strain rate and calculated the critical strain rate required in order to stabilize the anisotropic phase in a dilute solution.

Sluckin [27] adopted a quasi-equilibrium thermodynamic approach to understanding the effect of a strain rate field on the isotropic-anisotropic transition in polymer solutions. He derived a Clausius-Clapeyron-like equation which connects the shift in the critical polymer mole fraction C_I and C_N, which are concentrations of isotropic and nematic phases, respectively, to the applied strain rate.

It was shown by Helfrich [28] that at temperatures slightly above the nematic-isotropic transition temperature, a field induced phase transition from isotropic to nematic can be observed by applying an electric field. According to a thermodynamic consideration, the shift in the transition temperature caused by the electric field E is given by

$$\Delta T = E^2(T_0/q) (\varepsilon_2 - \varepsilon_1)/8\pi\varrho \tag{7}$$

where T_0 is the transition temperature in the absence of the electric field, q is the heat of transition per unit mass, ϱ is the density and ε_2 and ε_1 are the dielectric constants of the solution below and above the transition temperature. In most phase transition $(\varepsilon_2 - \varepsilon_1)/q$ is too small to detect a shift at practicable field strength.

The value of the dipole moment of PBLG molecule was determined to be 3.5 Debye per residue [29] which gives very large values such as 1750 Debye per molecule parallel to the long axis of the helix of PBLG with a degree of polymerization of 500. It can be expected that the effect of the electric field on the transition temperature is much larger in the rod-like polypeptide systems. Sukigara et al. [30] have reported the result of the shift of the transition temperature caused by an electric field on the system of PBLG in dioxane. According to their observation, when the temperature of an isotropic solution was decreased in the presence of an electric field, a homogeneous phase which transmitted light suddenly appeared at a certain temperature without passing through any biphasic system. The value of the critical temperature T_c was found to depend on the field strength and PBLG concentration. When T_c for PBLG solutions in dioxane at various concentrations is plotted against the square of the field strength, E^2, the plots fall on a straight line at a field strength lower than about 500 V/cm, in accord with Eq. (7) derived from the thermodynamic consideration of the phase transition. Furthermore they investigated the effect of an external electric field on the phase transition from isotropic to liquid crystalline phase in rod-like polymer solution, by applying the lattice model [31]. The free energy of each phase and the chemical potential of rod-like molecule and that of solvent in each phase were calculated. With these results a phase diagram was obtained. They concluded that application of an electric field resulted in narrowing the region of phase separation and shifting the phase transition concentration to a lower value. Thus, the change of transition concentration ΔC_c by an external electric field is given by,

$$\Delta C_c = - \frac{p^2 E^2}{l k^2 T^2} \tag{8}$$

where l is the axial ratio of a rod-like polymer, p is the dipole moment along its long axis. The plot of observed ΔC_c against E^2/T^2 shows a good linearity. The change in transition concentration is proportional to the square of an external electric field as is the change in the transition temperature.

2 Cholesteric Liquid Crystals in Polypeptide Solutions

2.1 Optical Properties of Cholesteric Liquid Crystals

In general, cholesteric liquid crystals are found in optically active (chiral) mesogenic materials. Nematic liquid crystals containing optically active compounds show cholesteric liquid crystalline behavior. Mixtures of right-handed and left-handed cholesteric liquid crystals at an adequate proportion give nematic liquid crystals. From these results cholesteric liquid crystals are sometimes classified into nematic liquid crystals as "twisted nematics". On the other hand, cholesteric liquid crystals form "bâtonnet" and terrace-like droplets on cooling from isotropic liquids. These behaviors are characteristic of smectic liquid crystals. Furthermore, cholesteric liquid crystals correspond to optically negative mono-axial crystals, different from nematic and smectic liquid crystals. Therefore, it is reasonable to regard cholesteric liquid crystals as the third type of liquid crystals.

Cholesteric liquid crystals are similar to smectic liquid crystals in that mesogenic molecules form layers. However, in the latter case molecules lie in two-dimensional layers with the long axes parallel to one another and perpendicular or at a uniform tilt angle to the plane of the layer. In the former molecules lie in a layer with one-dimensional nematic order and the direction of orientation of the molecules rotates by a small constant angle from one layer to the next. The displacement occurs about an axis of torsion, Z, which is normal to the planes. The distance between the two layers with molecular orientation differing by 360° is called the cholesteric pitch or simply the pitch. This model for the supermolecular structure in cholesteric liquid crystals was proposed by de Vries [32] in 1951 long after cholesteric liquid crystals had been discovered. All of the optical features of the cholesteric liquid crystals can be explained with the structure proposed by de Vries and are described below.

An examination of the polypeptide liquid crystal with a polarizing microscope reveals a number of remarkable features. The most prominent characteristic is a set of equally spaced parallel lines somewhat reminiscent of a fingerprint. If the polypeptide solutions are confined in a thin cell, the fingerprint pattern can not be observed, but a uniform area appears. This uniform area shows bright colors when observed with crossed polars and manifests a very high form optical rotation with a correspondingly high rotatory disperion. These experimental results are explained by a helicoidal structure shown in Fig. 6. In the drawing, the helical polypeptide molecules are represented by the rods rotating by a small constant angle from one layer to the next. When observed at right angles to the helicoidal (twist) axis, maxima in the major refractive index and birefringence will be observed on the layer where the molecules are at right angles to the direction of observation and minima where the molecules are parallel to the direction of observation, so that the periodicity S, obtained in microscopic observation as the distance between the alternating bright and dark lines, is equal half the pitch of the torsion. On the other hand, when observed in a direction parallel to the axis of torsion, no periodicities are seen, but a high degree of optical rotation is observed. The magnitude of S is not sensitive to the molecular weight of the polypeptide, but does depend on the polymer concentration, the solvent and the temperature. S may be as large as 100 μm or have values too small for resolution with an optical microscope. A very striking characteristic of cholesteric liquid crystals is their property of reflecting bright iridescent colors when illuminated with white light. When the spatial period of the cholesteric

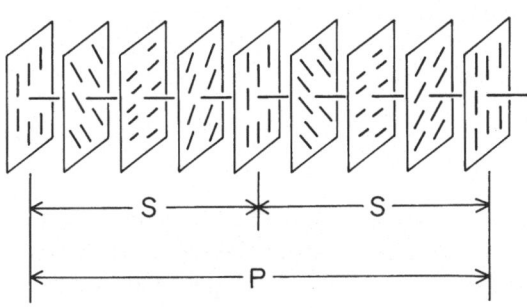

Fig. 6. Helicoidal structure of polypeptide liquid crystals

structure is of the order of the wavelength of light, the regular spacing acts as an internal diffraction grating. Thus, when white light is incident on the surface of the liquid crystal, selective reflection takes place over a small region of the spectrum, the wave length of maximum reflection varying with the angle of incidence in accordance with Bragg's law,

$$\lambda = 2\,Sn\,\sin\theta \tag{9}$$

where n is the refractive index of the solution, λ is the wavelength of the light and θ is the angle between the planes and the direction of the incident light. Iridescent colors had in fact been observed with solid films slowly evaporated from solutions of PBLG and poly(γ-methyl L-glutamate) (PMLG) [33, 34] and a solution of poly-(γ-ethyl L-glutamate) (PELG) in ethyl acetate [2] where very small pitches can be obtained in solutions.

Other equally remarkable optical properties are associated with the selective reflection. At normal incidence, the reflected light is circularly polarized; one circular component is totally reflected, while the other passes through unchanged. Also, quite contrary to what is found in normal substances, the reflected wave has the same sense of circular polarization as that of the incident wave. This is an important difference between the nature of the optical rotation of normal substances and of cholesteric liquid crystals. While the more familiar cases of optical rotation have their origin in the selective absorption of one circularly polarized component of the light, the "form" optical rotation of the twisted structure in cholesteric liquid crystals originates in the selective reflection of one circularly polarized component of the light.

The circular polarized dichroic ratio D is represented in the following equation.

$$D = (I_R - I_L)/(I_R + I_L) \tag{10}$$

where I_R and I_L are the optical intensities of the transmitted right-handed circularly polarized light and the left-handed one, respectively. Therefore, if cholesteric liquid crystals which reflect right-handed circularly polarized light and transmit the opposite handed light, are denoted as right-handed cholesteric liquid crystals, they manifest negative circular dichroism (CD) with a peak located at the wavelength of selective reflection.

The high optical rotatory power is observed in the uniform area of polypeptide liquid crystals, in which no retardation lines are present indicating that the structure is being viewed along the twist axis Z. De Vries derived an equation relating the optical rotatory power to the cholesteric pitch, the wavelength of the light and the birefringence. De Vries theory leads to the following formula for rotatory power Θ.

$$\Theta = \frac{\pi P}{4\lambda^2} \frac{(n_2 - n_1)^2}{1 - (\lambda/\lambda_0)^2} \tag{11}$$

where λ is the wavelength, P is the pitch, λ_0 is the wavelength of the selective reflection and $(n_2 - n_1)$ is the refractive index of the molecular layer. The sign of the pitch is positive for the right-handed cholesteric liquid crystal. When (λ/λ_0) is small compared with unity the formula reduces to

$$\Theta = \frac{\pi P}{4\lambda^2}(n_2 - n_1)^2 \text{ radian/micron}$$ (12)

$$\Theta = \frac{45(n_2 - n_1)^2 P}{\lambda^2} \text{ degree/micron}$$ (13)

In Fig. 7 the optical rotatory dispersion (ORD) as well as the circular dichroism (CD) is shown for the right-handed cholesteric liquid crystal. A right-handed helical structure reflects right circularly polarized light and it shows positive optical rotation on the short wavelength side of the reflection band.

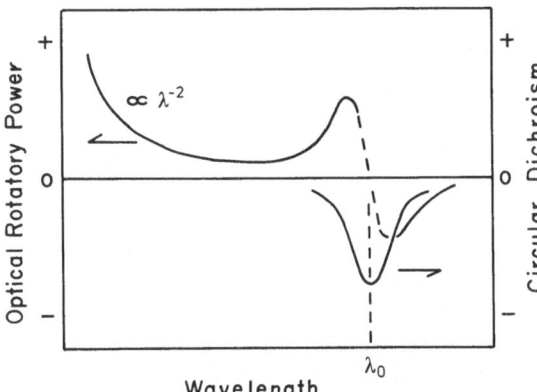

Fig. 7. Optical properties of right-handed cholesteric liquid crystals

Saeva and Wysocki [35] reported that achiral dye molecules such as acridine orange, anthracene, and pyrene become optically active when dissolved in a thermotropic cholesteric phase and a strong circular dichroism is induced in the wavelength region of the absorption bands of the dye molecules (liquid crystal induced CD, LCICD). Saeva also indicated that the induced circular dichroism of the achiral molecule in the thermotropic liquid crystal can be used as probe to determine the existence and the chirality of the cholesteric liquid crystalline phase. The achiral dyes dissolved in the concentrated solution of polypeptides show also the circular dichroism [36] and this will be induced by the dissymmetric field effect of the cholesteric helical structure on the electronic transitions of the dyes. The solute molecules dissolved in the mesophase preferentially orient their long molecular axes parallel to the alignment of the liquid crystal molecules. The principal axis of the solute molecule has a statistically preferred orientation relative to the director of the local nematic layer. The uniform angle of twist between successive nematic layers making up the cholesteric mesophase produces a helical array of the solute molecules. This structure induces circular dichroism. Sackmann and Voss [37], and Holzwarth and Holzwarth [38] explained the CD spectra in terms of the theory of light propagation in a cholesteric mesophase of de Vries, assuming the helical arrangement of the solute molecules. For this arrangement, the signs of the induced CD bands polarized parallel and perpendicular to the long axis

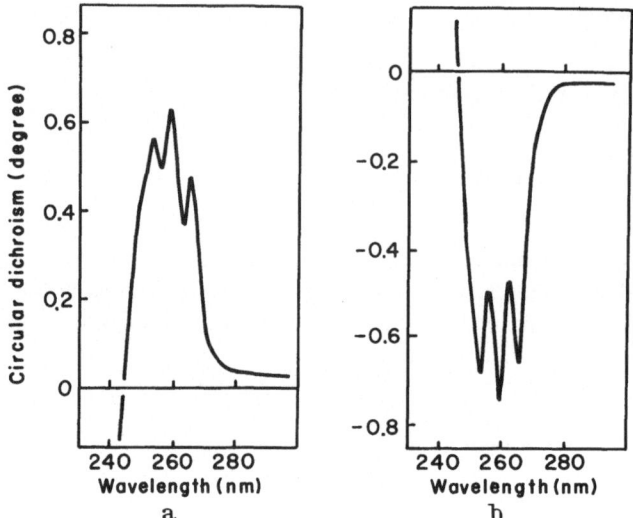

Fig. 8a and b. LCICD spectra of side-chain phenyl groups of PBLG in liquid crystalline solutions at room temperature; **(a)** c = 15.0 % in TCP and **(b)** c = 20.0 % in EDC

of the solute molecule are different from each other. Therefore the induced CD spectra have both positive and negative bands in the wavelength region of the absorption bands of the solute molecule. This is the case with thermotropic cholesteric mesophase. That is, in the thermotropic mesophase the sign of the pitch band CD is opposite to that of the long-axis polarized transitions of the solute molecules. However, the sign of the induced CD bands are of a single sign and opposite to that of the pitch band of the polypeptide mesophases. In the lyotropic mesophase, the solute-solvent interaction may be small compared with those in the thermotropic mesophase, so that the direction of the solute molecules in each of the thin layers is not uniform. As a result, the long- and short -axes polarized transitions of the solute molecules are equally exposed to the helical arrangement of the mesophase.

The terminal benzene rings of poly(γ-benzyl glutamate) are also exposed to the helical arrangement of the mesophase and the induced CD bands are observed in the wavelength region of the absorption of benzene rings.

Fig. 8a shows the CD spectrum of the liquid crystalline PBLG solution in 1,2,3-trichloropropane(TCP) at room temperature [39]. A positive circular dichroism in 240 ~ 270 nm can be assigned to LCICD of the side-chain phenyl groups of PBLG, since it has identical shape and peak wavelengths as those of an absorption spectrum and it disappears when the PBLG concentration is decreased to convert the liquid crystalline phase to the isotropic phase. This assignment is further supported by Fig. 8 b, where the spectrum changes its sign to negative in the left-handed cholesteric phase in EDC.

The influence of temperature on magnitude of the side-chain LCICD was examined in dioxane. An independent measurement under a polarizing microscope showed that the cholesteric half-pitch in this solvent increased with temperature. Therefore the LCICD magnitude was expected to increase according to the theoretical

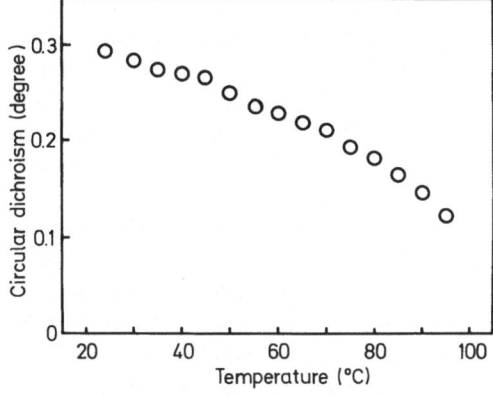

Fig. 9. Temperature dependence of the LCICD magnitude of the side chain phenyl band at 254 nm in the PBLG-dioxane (c = 20.0 %) liquid crystal

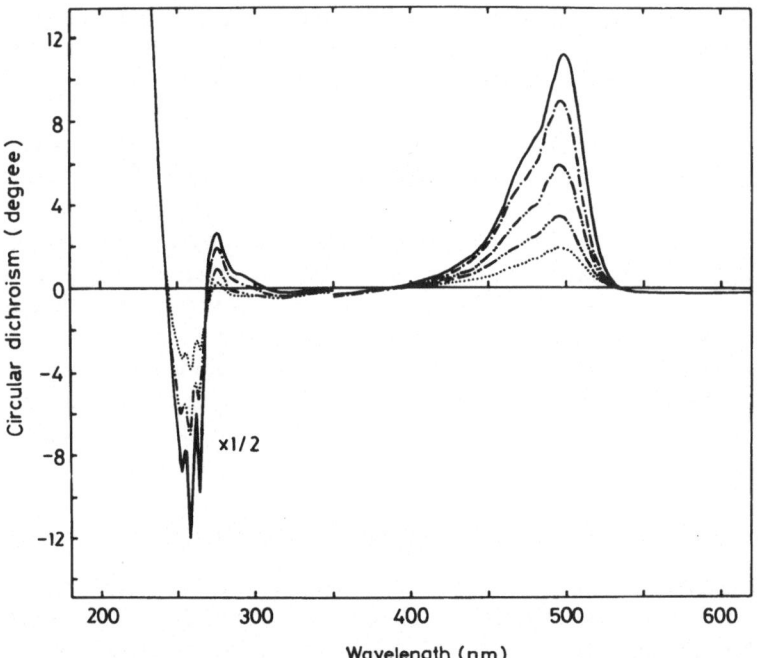

Fig. 10. LCICD spectra of a PBLG-EDC-acridine orange system at various temperatures. The polymer concentration is 15.1 % and temperatures are (—) 31 °C, (— · —) 42 °C, (— · · —) 50 °C, (— · · · —) 56 °C, and (· · · ·) 60 °C

prediction. The observed magnitude in Fig. 9, however, shows a monotonous decrease with increasing temperature. As indicated earlier by magnetic susceptibility measurements [40], a temperature increase reduces the degree of order of the liquid crystalline structure. This information allows us to suggest that a decrease in the LCICD magnitude at elevated temperatures is due to a randomization of the terminal benzene ring ordering, which dominates over the effect of an increase in the cholesteric pitch.

Fig. 10 shows the temperature dependence of LCICD for a PBLG-EDC-acridine orange system. There appears an intense positive band with a peak at 495 nm and a shoulder at 470 nm, and a relatively weak positive band in the $270 \sim 300$ nm region, in addition to the negative triplet induced on the side-chain phenyl groups. Since PBLG-EDC liquid crystals show no circular dichroism in the wavelength range longer than 270 nm, all positive LCICD bands are attributed to acridine orange. Linear dichroism studies [44] have shown that absorptions at $\lambda_{ab} = 495$ nm and 470 nm are associated with electronic transitions along the long, in-plane axis of monomeric and dimeric acridine orange, respectively. The $270 \sim 300$ nm band includes two transitions; parallel to the long ($\lambda_{ab} = 270$ nm) and the short ($\lambda_{ab} = 295$ nm) axes. However, observed LCICD spectra do not reflect these characteristics. They have a single sign and follow the shape of an absorption spectrum. We can conclude that LCICD signs for guest dyes are determined only by the cholesteric sense of the host matrix, independently of both the chemical structure and the orientation of the transition dipole moment within the dye molecule. The decrease in the magnitude with temperature is observed for every bands, and is again explained in terms of the reduction of the orientational order.

Recently Hashimoto et al. [42] proposed another method to determine the sense of cholesteric twisting, based on the distortion in the angular distribution of the depolarized laser light scattering patterns. In order to calculate the effect of the form-optical rotation on the scattering from the cholesteric twisted structure, they used the procedure by Picot and Stein [43] for calculating the effect on spherulitic scattering. They showed that if the optical rotatory power of the cholesteric liquid crystal, K, is positive, the intensity at $\mu = -45°$ (μ, the azimuthal angle) is greater than that at $\mu = +45°$ in the H_v pattern and if K is negative, the pattern is distorted in an opposite direction. Thus one can determine the sign of K and therefore the sense of the cholesteric twisting from the distortion of the pattern. It was also shown that the V_H pattern is identical to the pattern obtained by a 90° rotation of the H_V pattern around the incident beam axis.

2.2 Dependence of the Cholesteric Pitch on Temperature and Concentration

The temperature dependence of the cholesteric pitch in the polypeptide liquid crystals has been investigated in various solvents. The pitch P is related to the twisting angle φ between neighboring molecules separated by a distance d along the axis of torsion as follows.

$$\varphi = 2\pi d/P \tag{14}$$

Thus, the reciprocal value of P or S (half the pitch) provides a measure of the helical twisting power. The linear dependence of the helical twisting power on temperature is well explained by Keating's theory [44]. He proposed the theory of the cholesteric twist based on the hindered rotation of molecules due to thermal excitation in a plane normal to the axis of torsion. If we plot the variation of Keating's potential with φ along the horizontal axis, we get an asymmetrical curve around the vertical axis with a minimum at $\varphi = 0$. The time-average of the twisting angle does not vanish, because of the anharmonicity in the forces resisting the twist between neighboring planes of molecules. The macroscopic twist is treated as rotational analogue of thermal expansion with the dominant anharmonic forces coming from nearest neighbors. This theory provides a relationship between the average twisting angle $<\varphi>$ and the temperature,

$$<\varphi> = \frac{Ak}{2I\omega^4} T \tag{15}$$

where k is the Boltzmann constant, ω is the frequency of the excited twist mode, I is the moment of inertia of the molecule, and A is the constant in the cubic anharmonicity term of the anharmonic equation of motion. Equation (15) provides a linear dependence of $<\varphi>$ on T.

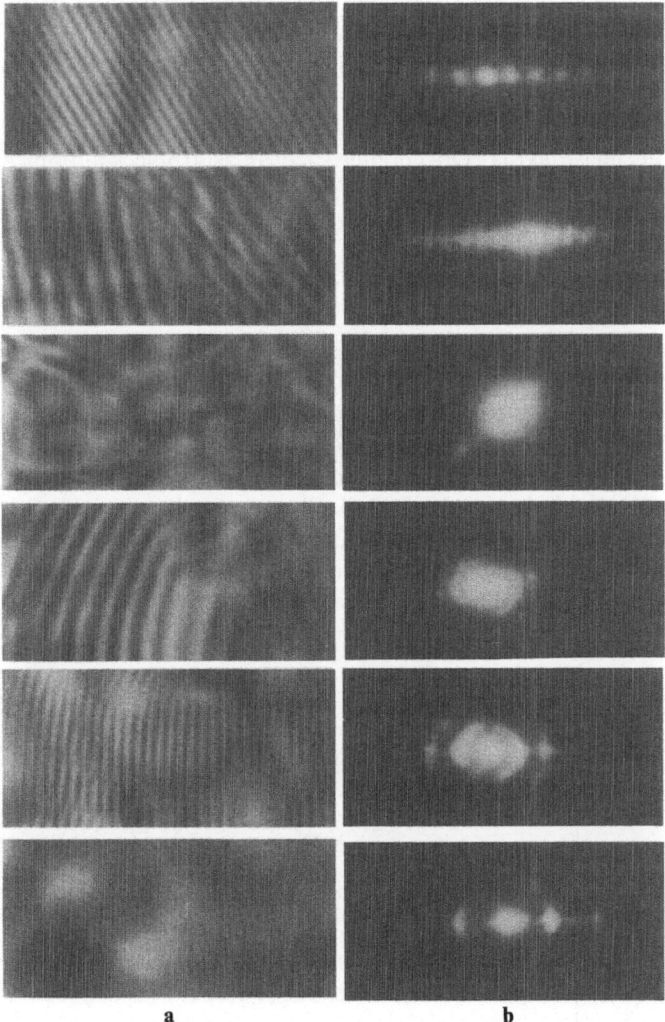

a   b

Fig. 11a and b. Cholesteric liquid crystalline structure of PBLG in m-cresol: **a**, striation patterns observed under a polarizing microscope; **b**, optical diffraction pattern with a beam from the He—Ne gas laser. Concentration is 17% volume fraction of polymer, and cell thickness is 2 mm

The linear relation holds quite well for a number of polypeptide mesophases [45]. The linear relation between 1/P and temperature suggests the existence of the nematic temperature T_N, where P tends to infinity and the mesophase becomes nematic. Actually the dark and bright striations observed by microscopy disappears and the thread-like structures characteristic of the nematic liquid crystal can be observed in the neighbourhood of T_N. T_N shifts to higher temperature with increasing concentration of polymer. Further investigation of the PBLG-m-cresol system [46] which makes it possible to extend the temperature range of the measurement widely, elucidated the reappearance of the dark and bright striations at temperatures above T_N. Fig. 11a shows a series of photographs of the striation patterns of the PBLG liquid crystals in m-cresol (17 vol % of PBLG) observed under a polarizing microscope. The distance between striations increases with temperature, starting from about 10 μm at 30 °C. At 60 °C, the striation patterns are replaced by threads characteristic of nematic liquid crystals. Above this temperature, the striation patterns appear again, and the spacing decreases with temperature. In the high temperature range (130~160 °C), a biphasic (isotropic, liquid crystalline) region appears. Above 160 °C, only the isotropic phase exists. On cooling such an isotropic solution, a birefringent phase appears first as spherulites with black maltase-crosses. The PBLG molecule has been shown to retain its α-helical conformation in m-cresol up to 200 °C. Therefore, this change should correspond to the phase transition in the narrow biphasic region in the temperature-composition phase diagram. Fig. 11b shows the optical diffraction patterns of the liquid crystalline solution of PBLG in m-cresol. A one-dimensional array of diffraction spots observed at lower temperatures indicates the presence of regular gratings in the liquid crystalline solution. The diffraction angle decreases with temperature in the range below 60 °C. The solution at 60 °C shows only a diffuse central spot. Above 60 °C, the diffraction angle increases with temperature, but the diffraction pattern changes its shape into a circular one. We obtain the value of the cholesteric half-pitch (S) using the Bragg's selective reflection formula. Optical rotator power Θ was measured for six different wavelengths (λ = 400, 500, 550, 600 and 650 nm) at several temperatures. By plotting Θ against $λ^{-2}$, straight lines were obtained. The sign of Θ changes from positive to

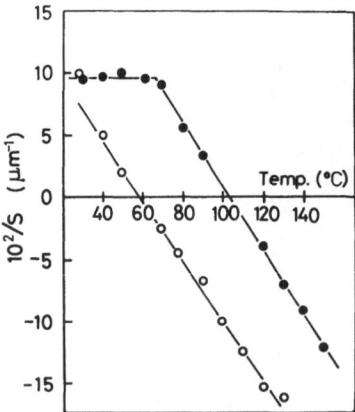

Fig. 12. Plots of the reciprocal half-pitch vs. temperature for PBLG liquid crystals: (○) in m-cresol; (●) in benzyl alcohol

negative at 60 °C. This is a direct proof for the inversion of the cholesteric sense from right to left. The solution at 60 °C shows no form optical rotation.

A similar thermally-induced inversion of the cholesteric sense was observed for the PBLG liquid crystal in benzyl alcohol. In this solution, a gel-like opaque phase coexists with the cholesteric phase at lower temperatures. The opaque phase disappears around 70 °C, where endothermic peaks are observed in the differential scanning calorimetry curve. The value of S below 70 °C remains constant, and then changes with temperature above 70 °C. The compensation occurs at about 103 °C, and the transition from biphasic phase to the isotropic phase is observed above 150 °C in this case. The results are summarized in Fig. 12, where the reciprocal of the half-pitch is plotted against temperature. The sign of 1/S is taken as positive when the cholesteric sense is the right-handed.

The temperature dependence of 1/S for poly(γ-methyl L-glutamate) (PMLG), poly(γ-ethyl L-glutamate) (PELG), and poly(γ-propyl L-glutamate) (PPLG) liquid crystals in m-cresol also shows linear relations. The PPLG liquid crystal exhibits a compensation at about 67 °C, and the cholesteric sense inverts from right to left. The cholesteric pitches of PMLG and PELG liquid crystals decrease monotonously with temperature. Extrapolation of straight lines indicates that the compensation should occur at 10 and 2 °C, respectively for PMLG and PELG liquid crystals. From these results it is concluded that if T_N is lower than room temperature, the twisting power increases linearly with temperature, while if T_N is higher than the temperature range of measurement, it should decrease with temperature. When T_N is in the temperature range of measurement, the inversion of the cholesteric sense can be observed.

The thermally induced inversion of the cholesteric sense was considered to be related to the intermolecular hydrogen-bond formation of the side chain ester groups of PBLG with m-cresol, since it was first observed in m-cresol solutions. However, our subsequent investigations [47] revealed that the sense inversion also occurs in 1,2,3-trichloropropane(TCP) which does not have any particular functional groups. A similar finding was recently reported in 1,1,2,2-tetrachloroethane [41].

The twisting power is an increasing function of concentration, as Robinson has shown that the S^{-1} is proportional to c^2 in dioxane. However, the relationship is only valid at a particular temperature and the relationship valid for any temperature is expressed as follows,

$$S^{-1} \propto c^n \tag{16}$$

where the exponent n varies with temperature in the range between 1 and 2. Taking account of the variation of the twisting power with temperature and concentration, the following equation has been proposed [44].

$$S^{-1} = a(1 - T/T_N) \tag{17}$$

$$a \propto c^n \tag{18}$$

The compensation temperature T_N is also a function of concentration [47].

According to Keating's theory, the pitch decreases with increasing temperature. However, the pitch in the cholesteric liquid crystals of polypeptides increases below T_N. Furthermore the cholesteric sense inverts at T_N. These results can not been explained by Keating's theory. What Keating provided is really a description of the solution to the problem. A truly microscopic theory should begin with an intermolecular force from which one derives the anharmonic mean field and thus the cholesteric phase. In 1970 Goossens [48] extended the molecular theory of Maier and Saupe [51] to take into account dipole-quadrupole as well as dipole-dipole interactions. The intermolecular potential obtained is given by

$$V(1, 2) = -(3/16d_{12}^4) [\alpha \cos 2\psi + (2\beta/d_{12}) \sin 2\psi] \qquad (19)$$

when ψ is the angle between the direction of the long axis alignment in respective planes 1 and 2 separated by a distance d_{12}. The coefficient of the symmetric part of this potential, α, is related to the anisotropy of the molecular polarizability and the coefficient of the asymmetric part, β, is related to the dispersion energy determined by dipole-quadrupol interactions. The sine term causes a shift of the potential minimum to a finite ψ, say ψ_0. By rewriting Eq. (19) as follows:

$$V(1, 2) = -(3/16d_{12}^4) \cos 2(\psi - \psi_0) \qquad (20)$$

where

$$2\psi_0 = \tan^{-1}(2\beta/\alpha d_{12}) , \qquad (21)$$

one finds a translated but symmetric potential curve. On account of the symmetry, temperature no longer has an effect on the average twist angle. Thus, although the Goossens model could give a proper magnitude of the pitch and succeeded in explaining the fact that the cholesteric structure appears only in optically active molecules, the temperature dependence of the pitch remained unexplained.

Lin-Liu et al. [49] indicated that the intermolecular potential is required to contain chiral contributions as a result of symmetry considerations, using a planar model and that, as temperature is varied, a particular pure cholesteric substance may undergo untwisting and reverse its helicity after passing through a divergence in pitch. Van der Meer et al. [50] published a molecular statistical theory based on the electric multipole expansion in analogy with Goossens' theory. They proposed an asymmetric potential and showed that the reciprocal pitch is approximately a linear function of temperature in agreement with esperiments.

Assuming the intermolecular force is the sum of a repulsion of hard-core with the shape of a twisted rod and of dispersion forces of Maier-Saupe-Goossens type, Kimura et al. [52] obtained a twisting power dependent on temperature, and explained the helix inversion with increasing temperature. Furthermore, they presented a statistical theory for a lyotropic cholesteric phase in the system of long coiled rod molecules [53], by applying the theory. They argued that the twisting power q is given as follows,

$$q = \frac{2\pi}{P} = \frac{24\lambda\Delta}{\pi LD} \, cf(c) \left(\frac{T_N}{T} - 1\right) \qquad (22)$$

where L and D is the length and the diameter of a rod, λ is a numerical factor ($\lambda \sim 1/2$ by Straley [54]), and Δ the height of the ridge of the coil which corresponds to the length of the side chain measured from the main cylindrical trunk of the polymer. f(c) is given by

$$f(c) = \left(1 - \frac{1}{3}\,c\right)\Big/(1 - c)^2 . \tag{23}$$

Experimental results can be explained quite well by the expression (22), which can be rewritten into the identical form as the empirical formula (17) at temperatures in which $(T - T_N) \ll T_N$. The constant a is given as

$$a = \frac{24\lambda\Delta}{\pi^2 LD}\,cf(c) \tag{24}$$

The results for five sorts of PBLG solutions are shown in Fig. 13, where L was assumed to be 1370 Å for the solute polymer with molecular weight 200,000, and the value of $\lambda\Delta/D$ adjusted independently of the concentration for each solvent. Throughout all cases, the theoretical curves reproduce the observed concentration dependence, and the estimates of $\lambda\Delta/D$ of the order 10^{-1} seem to be reasonable.

The dependence of a on the molecular weight is also well understood theoretically. The expression (24) indicates that the quantity a/c(f) is inversely proportional to the molecular mass. This relation is confirmed experimentally as shown in Fig. 14, where we find $\lambda\Delta/D = 0.13$. This is near to the value in the case (3) in Fig. 13.

T_N is given by the theory as follows,

$$T_N = \frac{\pi\tilde{B}RD}{4k_B L\lambda\Delta}\,\frac{1}{cf(c)} \tag{25}$$

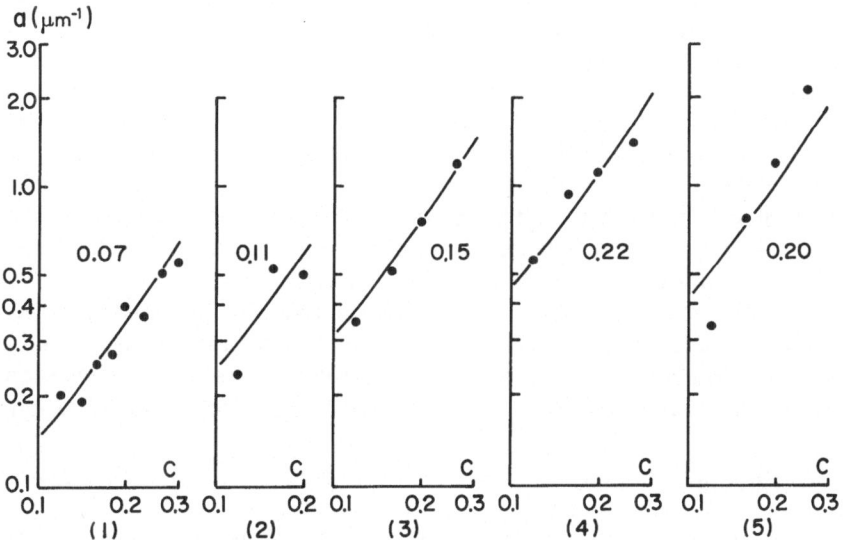

Fig. 13. Theoretical curves with experimental points of concentration dependences of the twisting power of PBLG (Mol. wt = 200,000) in various solvents; (1) $CHCl_3$, (2) CH_2Cl_2, (3) CH_2ClCH_2Cl, (4) $CHCl_2CHCl_2$, (5) $CH_2ClCHClCH_2Cl$, in which the number indicates the value of $\lambda\Delta/D$

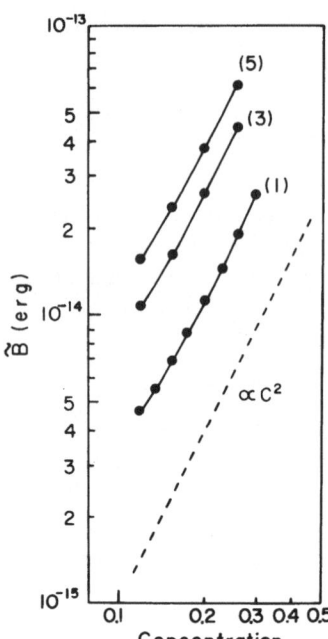

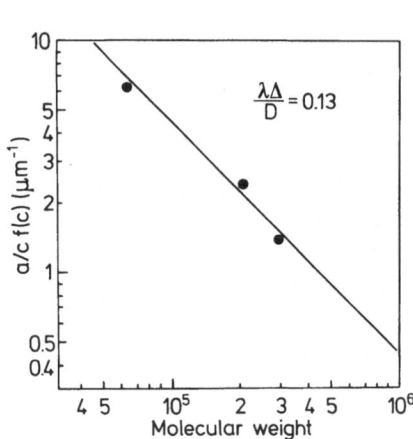

Fig. 14. Theoretical line with experimental points of the dependence of the twisting power of PBLG on the molecular weight in EDC. (c $\approx$ 0.20)

Fig. 15. Experimental concentration dependence of $\tilde{B}$ deduced from experimental data on T_N in various solvents: (1) $CHCl_3$, (3) CH_2ClCH_2Cl, (5) $CH_2ClCHClCH_2Cl$. The broken line gives the dependence proportional to c^2

where $\tilde{B}$ is a parameter relating to the attractive dispersion force proposed by Maier and Saupe and R is the mean distance between nearest neighboring molecules. The experimental data of T_N supply information on the quantity $\tilde{B}$. Those for PBLG with molecular weight 200,000 in three kinds of solvent were calculated. It enables us to calculate $\tilde{B}$ as a function of concentration, by making use of (25) with the numerical values of $\lambda\Delta/D$ given in Fig. 13. The results of calculation are shown in Fig. 15. It is noted that $\tilde{B} \propto c^2$ approximately. This dependence is expected for the Goossens model, in which the attractive potential is assumed to be $B(r) \propto r^{-7}$, although the validity of this dependence cannot be proved theoretically for such cases in which $R \lesssim L$ as in the present case of PBLG solutions.

Szarniecka and Samulski [55] have attributed the inversion of the twisting power occuring with the changes of temperature and of solvent to the transformation in the sense of coiling of the polymer. According to the theory by Kimura et al., the inversion can occur thermally, with no such essential transformation in molecular conformation.

2.3 Solvent Effects

In thermotropic liquid crystals, the cholesteric sense is determined by the chirality of the constituent molecule; the cholesteric sense of the liquid crystals of an optical isomer must be opposite to that of its mirror image isomer. When equal moles of both isomers are mixed (racemic mixture), the twisting power falls to zero and the cholesteric structure changes to a compensated nematic structure. Such compensation phenomena have been known for many binary mixtures of cholesteric materials or those of cholesteric and nematic materials. The first quantitative treatment of the pitch dependence on composition in binary mixtures was reported by Adams et al. [56]. Since then several approaches [57] to the helical interaction in a cholesteric liquid crystal have been used to explain qualitatively the composition dependence of the pitch in binary systems.

The compensation phenomena have also been known for lyotropic polypeptide liquid crystals in organic solvents. A racemic mixture of PBLG (a right-handed α-helix) and its mirror image isomer PBDG (left-handed α-helix) forms the nematic liquid crystal. However, a more surprising observation is that the cholesteric sense of the PBLG liquid crystal depends on the nature of the solvent. For example, PBLG liquid crystals in chloroform and in dioxane form right-handed cholesteric structure, while in methylene chloride and in ethylene dichloride(EDC), they form the left-handed one. In an appropriate solvent mixture of dioxane and methylene chloride (at 20 vol% of dioxane), the cholesteric structure changes to the compensated nematic structure.

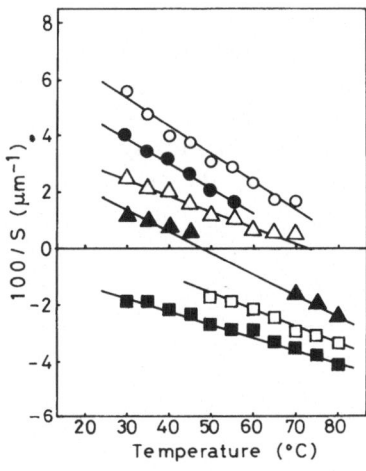

Fig. 16. Temperature dependence of the helical twisting power of PBLG liquid crystals in dioxane-EDC mixed solvents. (c = 0.15 vol%) The contents (vol%) ov EDC in solvent mixtures: (○) 0, (●) 20, (△) 40, (▲) 60, (□) 80, and (■) 100

Fig. 16 shows the temperature dependence of the reciprocal value of the cholesteric half-pitch of PBLG liquid crystals in dioxane-EDC mixed solvents (C = 15 vol%). [58] The sign of the S^{-1} value is taken as positive and negative for the right- and left-handed cholesteric structures, respectively. In dioxane, the right-handed cholesteric liquid crystal is formed and its twisting power decreases when temperature is increased. On the other hand, PBLG in EDC forms the left-handed cholesteric structure and the left-handed twisting power increases with temperature. Thus, the temperature coefficient of the twisting power is always negative for both systems. The addition of EDC to the dioxane solution causes a reduction of the right-handed twisting power, and produces a downward shift of the S^{-1} vs. temperature curve. In a mixture of 40 vol% dioxane and 60 vol% EDC, the liquid crystal undergoes a thermally-induced sense inversion from right to left, at about 47 °C. In all of these systems, a linear relationship between the twisting power and temperature holds.

In Fig. 17, the twisting power is plotted against the solvent composition in the dioxane-EDC mixed solvent system at various temperatures. It is clear that a linear additive law for the twisting power is realised at each temperature. The critical composition of the solvent at which compensation occurs shifts to higher dioxane content as the temperature is increased.

The temperature dependence of the twisting power in the m-cresol-EDC system has been also investigated. The cholesteric sense of the PBLG liquid crystal in m-cresol inverts from right to left as the temperature is raised above 60 °C. The addition of EDC to this system causes a shift of the compensation temperature to 37 °C at 20 vol% EDC and to below 20 °C at 40 vol% EDC. At the same time, the temperature coefficient of the twisting power gradually decreases. The downward shift of the S^{-1} vs. temperature curve continues up to 60 vol% EDC, but a further addition of EDC causes an upward shift. This suggests a quadratic dependence of the twisting power on the solvent composition. This can be clearly recognized in Fig. 18, where the twisting power is plotted against the solvent composition. When the EDC content is increased at low temperatures, the right-handed twisting power rapidly decreases to zero and inverts to the left-handed twisting power. In the higher EDC content region, the left-handed twisting power begins to decrease after going beyond an extreme value. The deviation of the experimental curve from the linear additive law seems largest at about 50% EDC. The solvent-induced compensation disappears above 60 °C, but the twisting power still maintains its quadratic dependence on the solvent composition.

More complicated behavior was observed in the m-cresol-dioxane system. The twisting power vs. the solvent composition curve shows both the maximum and

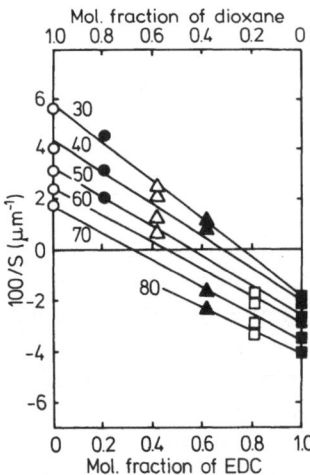

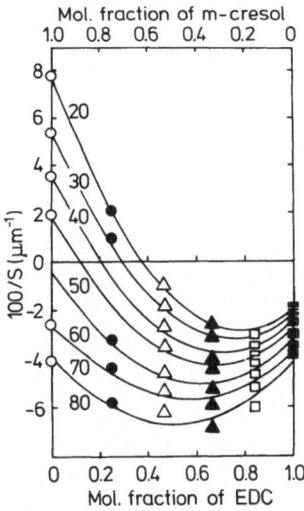

Fig. 17. Dependence of the helical twisting power on the solvent composition in dioxane-EDC mixtures. Numerals indicate the temperature in °C

Fig. 18. Dependence of the helical twisting power on the solvent composition in m-cresol-EDC mixtures

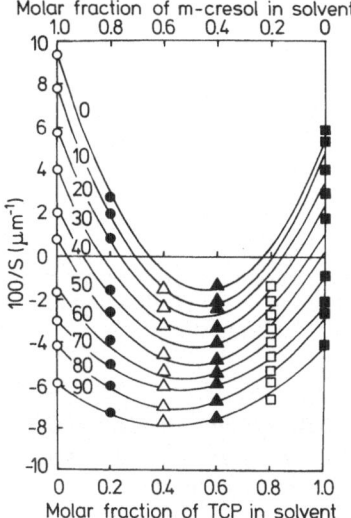

Molar fraction of m-cresol in solvent

Fig. 19. Dependence of the helical twisting power on the solvent composition in TCP-m-cresol mixtures

minimum peaks at low temperatures. The initial increase in the right-handed twisting power caused by the addition of m-cresol might arise from a similar mechanism bringing about an increase in the left-handed twisting power of the PBLG-EDC.

Fig. 19 illustrates the change of S^{-1} with the solvent composition in the TCP-m-cresol system. At low temperatures, a right $\rightarrow$ left $\rightarrow$ right sense inversion can be induced by increasing the TCP content. The twofold inversion disappears above 60 °C since the cholesteric structure is left-handed in both TCP and m-cresol.

It is known that α-helices of PBLG aggregate in certain solvents and that the structure of this aggregate depends on the solvent [59]. In the EDC solution, a linear head-to-tail type aggregation has been suggested for PBLG [60]. The addition of small amounts of certain agents (DMF, dichloroacetic acid (DCA), and trifluoroacetic acid (TFA)) to the system causes a disruption of the aggregated PBLG helices, as indicated by a decrease in the viscosity and by a change in the dipole moment.

When a few percent of TFA (up to 6%) were added to the PBLG-EDC liquid crystal, the fluidity of the solution was increased markedly. This behavior is similar to that in dilute solutions, thus indicating the breakdown of the aggregates. A further addition of TFA (8%) results in the phase separation of a small portion of an isotropic phase from the liquid-crystalline phase. According to the Flory theory of the phase separation, the critical concentration of polymer necessary to form the liquid-crystalline phase is expected to shift to higher values when the axial ratio (length-to-width ratio) of polymer is decreased or the flexibility is increased. Thus, this observation can be ascribed to the disruption of the linear head-to-tail type aggregation and/or to the increase in the flexibility of PBLG α-helices. The effects of TFA on the helical twisting power of the PBLG-EDC liquid crystal are shown in Fig. 20. The left-handed twisting power gradually increases with the TFA content. At the same time, the temperature coefficient of the twisting power increases. A similar observation was reported by DuPré et al. [61] for the PBLG-dioxane liquid crystal, wherein the right-

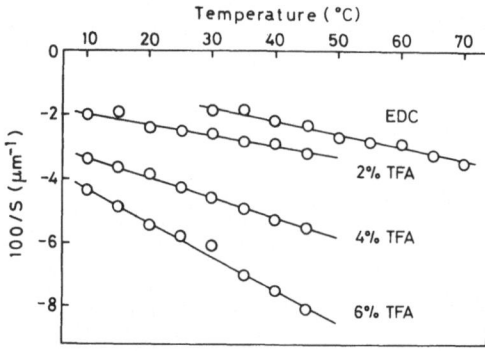

Fig. 20. Temperature dependence of the helical twisting power of PBLG liquid crystals in EDC-TFA mixtures

handed twisting power increases with TFA content. In both systems, disruption of the aggregates is accompanied by an increase in the twisting power of the starting liquid crystal, regardless of the cholesteric sense. In contrast to this, Patel and DuPré [62] found that the effect of TFA in the PBLG-benzene system is quite different from that observed in the cases of dioxane and EDC. The pitch was found to increase with increasing acid content. The difference in action of halogenated acids in these two systems is most probably a result of different specific polymer-solvent and solvent-solvent interactions. The physical properties of benzene and dioxane are similar (dielectric constants at 25 °C of 2.274 and 2.209, respectively), so that we must look for additional factors to understand the variance. It is known that small amounts of TFA protonate the N-terminal amide group of PBLG [63] and associate with the polymer by H-bonding to the carboxyl groups of the side chain ester [64]. The result is a reduction in the overall dipole moment of the helical macromolecule and a modification of the polarizability of local segments. Such modification of the polarizability of local segments. Such modification of the polymer would occur by acid attack in both systems. However, a strong acid such as TFA or DCA reacts with dioxane to form a solvated proton and the acid anion. Hydrocarbon solvents such as benzene do not form such an ion. The shift of acid molecules from polymer to solvent will be sufficient to reverse the effect of TFA on the twisting power.

Finkelmann and Stegemeyer [65] have analyzed the composition dependence of the twisting power in binary mixtures of thermotropic liquid crystals, using the Goossens theory. The twisting angle φ between the directors of two nematic layers is given as follows for a mixture of molecules of type 1 and 2,

$$\varphi = \frac{1}{d} \frac{B_{11}x^2 + (B_{12} + B_{21})\, x(1 - x) + B_{22}(1 - x)^2}{A_{11}x^2 + 2A_{12}x\,(1 - x) + A_{22}(1 - x)^2} \tag{26}$$

where d is the distance betwenn two directors and x is the molar fraction of component 1. The parameter A_{ij} is related to the anisotropy of the molecular polarizability, and B_{ij} is related to the dispersion energy determined by dipole-quadropole interactions. A non-vanishing B_{ij} value, responsible for the cholesteric twist, is attained only for optically active molecules. In the case of binary mixtures of cholesteric molecules having chemically similar properties, one would expect $A_{11} \approx A_{22} \approx A_{12} = A$. Then, by substituting $\varphi_{ij} = B_{ij}/dA$, Eq. (26) can be reduced to,

$$\varphi = \varphi_{11}x^2 + (\varphi_{12} + \varphi_{21})\, x(1-x) + \varphi_{22}(1-x)^2 \qquad (27)$$

further, by using the relationship $\varphi = \pi d/S$, we get

$$S^{-1} = S_{11}^{-1}x^2 + (S_{12}^{-1} + S_{21}^{-1})\, x(1-x) + S_{22}^{-1}(1-x)^2 \qquad (28)$$

Thus, the twisting power in binary mixtures is characterized by a quadratic dependence on the composition and by the appearance of the cross terms S_{12}^{-1} and S_{21}^{-1} representing the induced twisting power. If molecules 1 and 2 in a mixture satisfy the conditions $B_{21} \approx B_{11}$ and $B_{12} \approx B_{22}$, Eq. (28) may be further reduced to,

$$S^{-1} = S_{11}^{-1}x + S_{22}^{-1}(1-x) \qquad (29)$$

This equation provides a twisting power which varies linearly with composition. When S_{11}^{-1} and S_{22}^{-1} have opposite signs, the inversion of the cholesteric sense should occur at a critical composition, $x^* = S_{22}^{-1}/(S_{22}^{-1} - S_{11}^{-1})$.

The dependence of the twisting power of PBLG liquid crystals on the solvent composition is well described by the extended Goossens theory, provided we use x as the molar fraction of one solvent component in a solvent mixtures. The twisting power in the dioxane-EDC system varies linearly with the solvent composition, and it can be fitted to Eq. (29). On the other hand, in the m-cresol-EDC system, the twisting power has a quadratic nature, as indicated by Eq. (28).

The substitution of dioxane by m-cresol results in a remarkable influence on the twisting power. The deviation from the linear additive law observed in the m-cresol-EDC system indicates that the effect of adding m-cresol increases the left-handed twisting power of the PBLG-EDC liquid crystal. Similarly, by adding m-cresol to the dioxane solution, the right-handed twisting power increases. It is interesting to compare these observations with the results in the EDC-TFA mixed solvent; a trace amount of TFA increases the left-handed twisting power of the PBLG-EDC liquid crystal. The IR spectra of the carbonyl ester groups of PBLG in liquid-crystalline solutions in EDC, dioxane, and m-cresol has been investigated. The wavenumbers of maximum absorbance are 1735 cm^{-1} in EDC and dioxane, but 1715 cm^{-1} in m-cresol. Peaks at 1735 and 1715 cm^{-1} have been assigned to the carbonyl groups in the free and in the hydrogen-bonded state, [66] respectively. It appears that the carbonyl groups in the m-cresol solution are hydrogen-bonded with solvent molecules, thereby differing from those in the other two solvents. m-cresol added to the EDC or dioxane solution, breaks down the aggregates of the PBLG α-helices. The initial increases in the twisting power caused by m-cresol or TFA may be attributed to the disruption of the aggregates.

Samulski and Samulski [67] have presented a discussion on the solvent effect in terms of relative magnitudes of the dielectric constants of PBLG and solvent molecules. This theory predicts that compensations occurs for a critical value of the dielectric constant of solvent $\varepsilon^* = (\varepsilon_1 \varepsilon_2)^{1/2}$, where ε_1 and ε_2 are the principal values of the dielectric constant perpendicular to the long molecular axis of PBLG. This

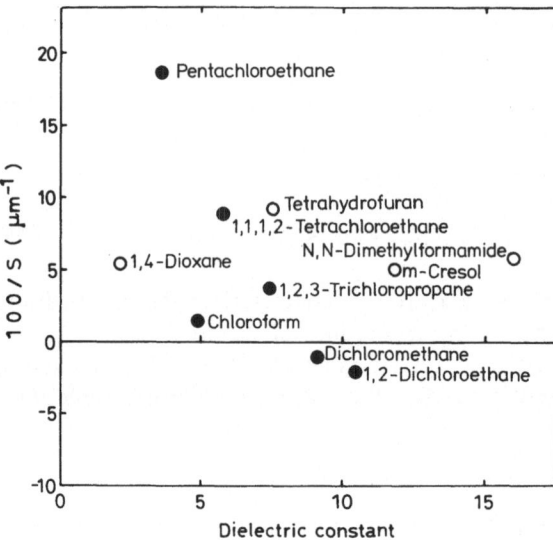

Fig. 21. Plots of the helical twisting power vs. dielectric constants of solvents at 25 °C

indicates that the twisting power changes its sign when the dielectric constant of the solvent is varied on passing through ε* by the addition of second solvent component. This theory seems to be true for the inversion phenomena in the dioxane-EDC system, where the right-handed cholesteric structure in dioxane (ε = 2.2) changes to the left-handed structure in EDC (ε = 10.5). With regard to the dielectric constant, one would expect m-cresol (ε = 11.8) to behave as EDC, i.e., m-cresol supports the left-handed cholesteric structure. However, the aforementioned study shows that m-cresol supports the right-handed structure at low temperatures. This discrepancy may again be ascribed to the short-range interactions of m-cresol with the PBLG side chains. Milstien and Charney [64] have noted that the formation of intermolecular hydrogen bonds changes the orientation of permanent dipole moments located on the carbonyl ester groups. This may be seen by a reduction of the overall dipole moment of the PBLG molecule with the addition of TFA. If such a change in the side-chain environment alters the anisotropy in the dielectric constants of PBLG (ε_1 and ε_2), then in accordance with the theory of Samulski and Samulski, we may expect the magnitude and sign of the twisting power in the m-cresol solution to be affected. It seems that the formation of right-handed cholesteric structure in benzyl alcohol (ε = 11.0) may also be explained by these mechanism.

The best agreement between theory and experiment will be achieved in a series of non-interactive, chemically similar solvents. Moreover, if a number of such solvents are chosen to cover a proper range of ε, the theory can be simply tested without preparing mixed solvents. The result of such a study [68] is shown in Fig. 21, where S^{-1} is plotted against ε. We find a good correlation between S^{-1} and ε for chlorinated hydrocarbons (shown by closed circles). The twisting power decreases with increasing ε and becomes negative for dichloroethane. The experimental value for chloroform shows a displacement from the expected point. However it still preserves a positive value. The sense inversion occurs around ε ≈ 9 and this value agrees with the calculated result of Samulski and Samulski (ε = 7.9) for the methylene chloride-

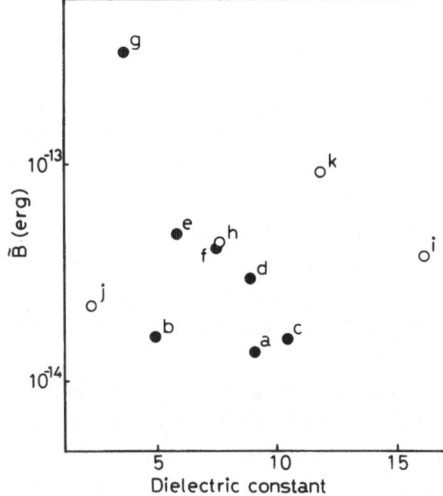

Fig. 22. Plots of the parameters $\tilde{B}$ vs. dielectric constants of solvents at 25 °C. a: dichloromethane, b: chloroform c: 1,2-dichloroethane, d: 1,2-dichloropropane, e: 1,1,1,1,2-tetrachloroethane, f: 1,2,3-trichloropropane, g: pentachloroethane, h: tetrahydrofuran, i: N,N-dimethylformamide, j: 1,4-dioxane, k: m-cresol

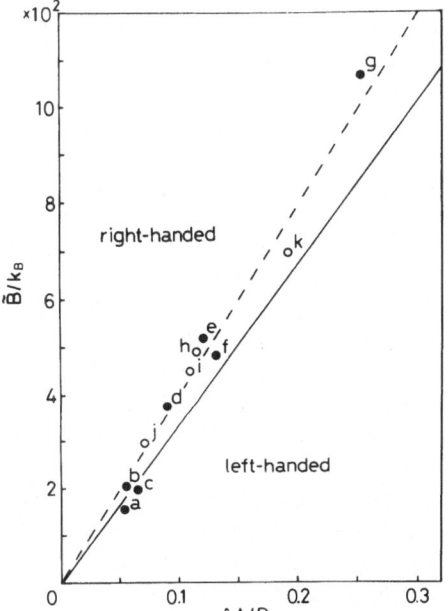

Fig. 23. Relation between $\tilde{B}/k_B$ and $\lambda\Delta/D$ in various solvents (c = 0.15 vol%). Letters in the figure are the same as those in Fig. 22

dioxane system. These observations suggest that the dielectric constant of the solvent plays an inportant role in determining the helical twisting power in this particular series of solvents.

The analysis of the experimental results by the theory of Kimura et al. give interesting clues for understanding the polypeptide liquid crystal. In Fig. 22 the parameters $\tilde{B}$, which are related to the attractive dispersion force proposed by Maier and Saupe and can be obtained from the slopes in the $1/S$ vs. $1/T$ plots or the values of T_N, are plotted against the dielectric constants of solvents. $\tilde{B}$ is expected to decrease with increasing dielectric constant. This seems to be valid for a series of chlorinated hydrocarbons (shown as closed circles) with the exception of chloroform.

In Fig. 23 the relationship between $\tilde{B}/k_B$ and $\lambda\Delta/D$ are shown for the polypeptide liquid crystals in various solvents. Following Eq. (25), the compensation temperature T_N is determined by the ratio $(\tilde{B}/k_B)/(\lambda\Delta/D)$ at a constant polymer concentration. The solid and broken lines shown in Fig. 23 correspond to the theoretical values calculated for $T_N = 25$ °C and 80 °C, respectively. The solvents located above and below the solid line support the right-handed and left-handed cholesteric liquid crystals, at 25 °C, respectively. The situation is the same for the broken line at 80 °C. The solvents located between the two straight lines invert the cholesteric sense from right-handed to left-handed in the range of measurements.

The values of $(\lambda\Delta/D)$ and $\tilde{B}$ obtained from the temperature dependence of the pitch in the EDC-dioxane system show a linear change with the solvent composition in the same manner as the twisting power. For the EDC-m-cresol system which shows a quadratic variation of the twisting power with the solvent composition, a quadratic variation in $\lambda\Delta/D$ and $\tilde{B}$ is also observed. As $\tilde{B}$ is related to the dielectric constant, the variation of the twisting power with solvent composition is related to the change of the dielectric constant with solvent composition in mixed solvents. A linear change of the dielectric constant will occur in the EDC-dioxane system, while a quadratic change will be expected in the EDC-m-cresol system.

3 Cholesteric Twisted Structure in Solid Films of Polypeptides

In consideration of the concentration dependence of the cholesteric pitch, it is easy to envisage that a similar structure with a small pitch is retained in the solid films when all the solvent is removed. If the films have a cholesteric pitch comparable to the wavelength of visible light, it is expected from de Vries' theory, that they will show the cholesteric color and the form circular dichroism. Such observations have been reported for the films of poly(γ-methyl glutamate, PMG) and PBLG. Tobolsky et al. [69] have reported that, when solid films of PBLG are cast from solvents such as $CHCl_3$ or CH_2Cl_2, X-ray evidence and anisotropic swelling characteristics clearly indicate that the rod-like PBLG molecules lie in the plane of the film, but the helix axes of PBLG molecules are randomly oriented in this plane. These observations led them to intuit that these cast films retain the cholesteric structure found in the fluid liquid crystalline solutions of PBLG. Carefully prepared plasticized films, which are cast from mixed solvents (volatile solvent + plasticizer), exhibit the optical retardation lines characteristic of the cholesteric structure in the liquid crystalline phase of PBLG.

Tachibana and Oda [33] have performed circular dichroic studies with colored films of PMDG and PMLG. The CD curves exhibited a large band with a peak in the region $450 \sim 650$ nm which is attributable to selective reflection of circularly polarized light of one sense in that wavelength region. The PMDG films reflected selectively left circularly polarized light. On the other hand, the PMLG films were found to give the CD curve of the opposite sense. It is thus concluded that the PMDG films have a left-handed macrohelix, while the PMLG films have a right-handed one. Since the

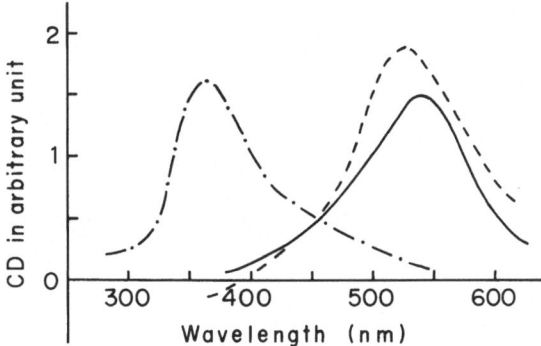

Fig. 24. Circular dichroism of PMDG films cast from EDC solution (—), from DCM-DMF solution (— —), and from chlroform-DMF solution (— · —)

molecular helix (α-helix) of PMDG (or PMLG) is known to be left-handed (or right-handed), this result indicates that the sense of the macrohelix is the same as that of the molecular helix of each enantiomer.

A colored film of PMG can be prepared for reproduction from three different solutions [34]. One of the films is cast from EDC solution, as reported by Tachibana and Oda. The other films are cast from dichloromethane(DCM)-DMF and chloroform-DMF solutions; 1 g PMDG is dissolved in 30 ml DCM (or chloroform) and then 0.4 ml DMF is added. Colorless films are prepared from chloroform or DCM solution, but from X-ray measurements they do not to contain the cholesteric twisted structure. Therefore, DMF seems to play an important role in the preparation of colored films. When observed at right angles to the film surface, the films cast from EDC and DCM-DMF solutions showed green or yellow color, while blue or violet colored films are prepared from chloroform-DMF solution. Fig. 24 shows the CD curve, where the CD is expressed as the apparent absorbance for the left circularly polarized light minus that for the right circularly polarized light. The CD curves for the green colored films from EDC and DCM-DMF solutions exhibit a large positive band with a peak in the region $500 \sim 600$ nm. For the violet colored film from the chloroform-DMF solution, a peak is observed in the $300 \sim 400$ nm. The values of the wavelength λ_m at peaks in the CD curves are in good agreement with the wavelength of visible color observed at right angles to the film surface. Thus, the colors of the films are due to the selective reflection of visible light. The large positive bands in $300 \sim 600$ nm further show that the reflected light is left circularly polarized.

It has been predicted that the range of total reflection should expand from $P(n - \Delta n/2)$ to $P(n + \Delta n/2)$ [70], where P is the pitch, n is the average refractive index and Δn is the birefringence of untwisted material. Therefore, the width of the reflection is given by the equation

$$\Delta \lambda = P \Delta n \quad \text{or} \quad \Delta \lambda / \lambda_m = \Delta n / n \tag{30}$$

From Fig. 24,

$$\Delta n / n = \Delta \lambda / \lambda_m = 0.20 \sim 0.25$$

Using $n = 1.4$, Δn is estimated as about 0.3 which is of the different order from the birefringence ($\Delta n = 0.026$) measured for the nematic phase in the solution which

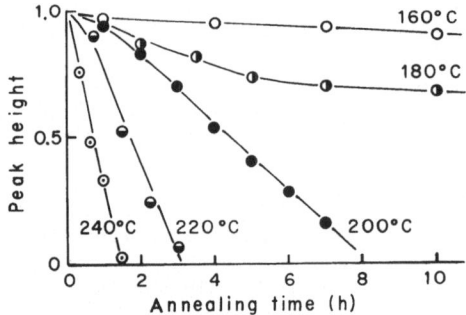

Fig. 25. Plots of the peak heigt in the CD curves against the annealing time for PMDG films (cast from EDC solution) annealed at various temperatures

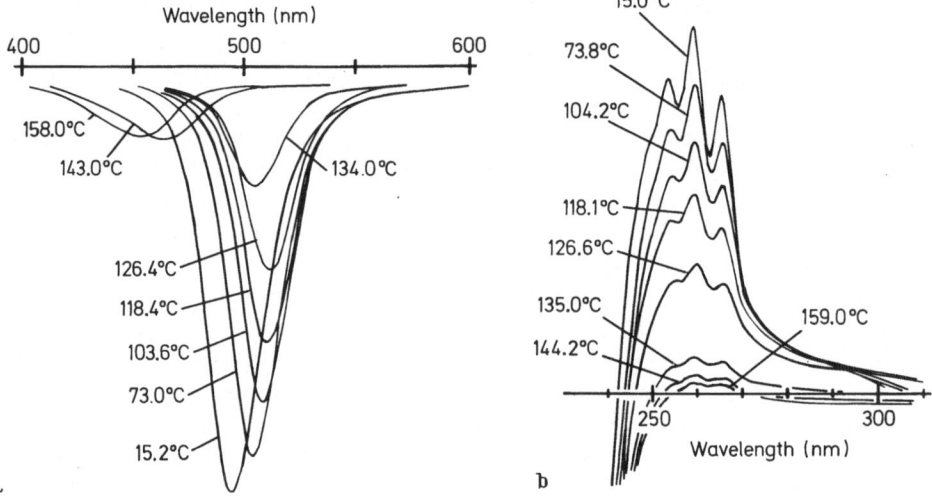

a   b

Fig. 26a and b. Temperature dependence of CD spectra of a colored PBLG film: **a**: cholesteric pitch bands around 500 nm; **b**: LCICD of the side-chain phenyl groups

contained equal quantities of the L and D enantiomorphs of poly(γ-benzyl glutamate). This discrepancy is probably caused by the wide distribution of pitch length and/or the disordering of the twisted axes.

The cholesteric twisted structure in the solid state is so stable on heating that it is not destroyed at temperatures below 180 °C. Above 200 °C, however, this cholesteric twisted structure is gradually destroyed depending on the heating duration. In Fig. 25, the normalized peak heights in the CD curves are plotted against the annealing time at various temperatures. The peak height decreased with heat application and reached zero after 3 hr at ·220 °C or 1 hr at 240 °C. This destruction of the cholesteric twisted structure is reflected in the X-ray diffraction patterns. The characteristic pattern of the colored films gradually changed to the hexagonal pattern by annealing above 200 °C. The film annealed at 240 °C for 1 hr, clearly showed the same hexagonal pattern as in the film cast from the chloroform solution. The results show that the destruction of the solid cholesteric structure is caused by the molecular rearrangement from the twisted array of α-helices to the parallel one and also, that the anharmonic molecular motion of the individual twisted α-helices may occur above 200 °C.

Colored films of PBLG are also obtained by casting from DMF-m-cresol mixed solvents (100:1) at 60 °C [71]. These films belong to form A, since the characteristic transition is observed around 135 °C. In Fig. 26a temperature dependence of CD spectra of a colored PBLG film is shown. The cholesteric pitch bands observed around 500 nm shift to lower wavelengths irreversibly above 135 °C, where the benzene ring stacking in the side chain is considered to break down. In Fig. 26b positive CD spectra observed in the region 240 ~ 270 nm, which are assigned to LCICD of the side-chain phenyl groups of PBLG in the solid film, are shown at various temperatures. The peak height decreases abruptly at about 135 °C.

Studies on thermotropically mesomorphic polymers have remarkably increased recently. They have usually been classified into two categories [72]. One is comb-like polymers which have mesogenic side chains attached to the main chain. The other type of polymer has rigid mesogenic units in the main chain. A twisted structure has often remained in the solid film of polypeptides, as described above. However, thermotropic cholesteric mesophase has not been found so far for polypeptides. Dimarzio [5] pointed out in his calculation that solvent molecules in lyotropic systems could be replaced by flexible polymers. This may be achieved by attaching long side chains to the α-helical main chain. For this purpose, we synthesized a series of copoly-(γ-n-alkyl L-glutamate)s having long side chains, and examined the thermotropic liquid crystalline behavior [73].

Three types of copoly(γ-n-alkyl L-glutamate)s with the combination of methyl-hexyl (MH), methyl-octyl (MO) and propyl-octyl (PO) were prepared by alcoholysis of PMLG in ethylene dichloride using p-toluenesulfonic acid as the catalyst at 60 °C. The thermotropic mesophasc was detected by the iridescent cholesteric color which appeared by annealing at temperatures in the region 110 ~ 190 °C. Also the circular dichroism due to the selective reflection was measured at room temperature for the quenched films. Only the films with the copolymer composition of about 50/50

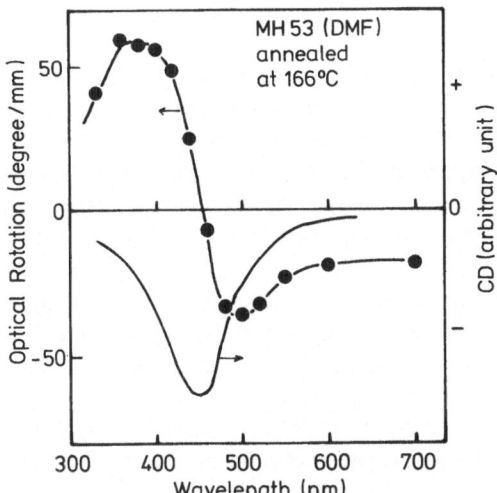

Fig. 27. ORD and CD spectra of a MH-53 film annealed at 166 °C

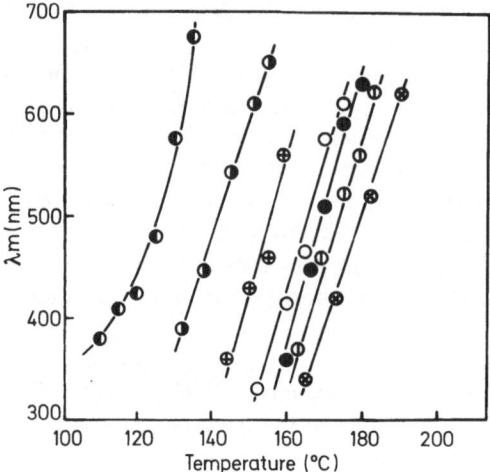

Fig. 28. Plots of λ_m vs. annealing temperature: (◯) MO-80, (◯) PO-63, (⊕) PO-60, (◯) MO-60, (◯) MH-58, (●) MH-53, (⊗) MH-50

in mol% showed the cholesteric colors when they were annealed at appropriate temperatures. Fig. 27 shows ORD and CD spectra measured for the MH-53 film, where the number following to MH denotes the mol% of the longer side chain. This optical property is characteristic of the right-handed twisted structure. All the other colored films exhibited the negative CD band. The peak wavelength of the CD band, λ_m, is plotted against the annealing temperature in Fig. 28. The range of the annealing temperature to produce the cholesteric color is limited within an interval of about 30 °C for each sample, and it is lower for the copolymer which has the larger amount of long side chains. The λ_m was independent of the type of casting solvent. For the initial 1 hr in the annealing process, the CD band shifted toward long wavelength and became sharp. An annealing time of 2 hr was generally enough for the equilibration. The λ_m position was reversible with regard to the change of the annealing temperature, although it required about 2 hr for complete reversibility. The peak wavelengths of CD increased with the annealing temperature. This shows the temperature dependence of the cholesteric pitch in copolymer films is the same as that in the polypeptide solutions.

The crystallinity of the mesomorphic copolymers was extremely low as judged from the X-ray diffraction pattern, in which only one broad reflection was observed. This may be an essential point with regard to the chain mobility. Since the long side chains are considered to play a role as solvent molecules, the thermotropic system may correspond to a highly concentrated solution. α-Helical main chains embedded in the matrix of side chains take the cholesteric arrangement because of the chirality of the main chain.

Gray et al. [74] have reported that (acetoxypropyl)cellulose behaves as a thermotropic cholesteric liquid crystal below 164 °C. It has been also observed that some (hydroxypropyl)cellulose forms a thermotropic liquid crystalline phase at temperatures above 160 °C [75]. From these results together with our finding, we presume that rigid rod-like polymers having dissimilar and/or high density of flexible side chains may form thermotropic liquid crystals, in general. These polymers should be classified as the third category of thermotropically mesomorphic polymers.

4 Concluding Remarks

We have discussed in detail the cholesteric pitch and the cholesteric sense in polypeptide liquid crystals, centering around our own work, which had not been investigated much since Robinson had presented excellent works about a quarter of a century ago. We have show that compensation of the cholesteric sense caused by temperature and solvent can be consistently explained by the theory of Kimura et al., in which the intermolecular force is assumed as the sum of a repulsion of a hardcore with shape of a twisted rod and of dispersion forces of Maier-Saupe-Goossens type.

We have also discussed gelation in wide biphasic regions in the phase diagram. This occurs because the isotropic and anisotropic phases do not coexist in the wide biphasic region as theories predict, but actually gelation takes place. Formation of the "complex phase" is related to gelation and especially in the PBLG-DMF system, where benzene ring stacking in the side chains plays an important role. Therefore the PBLG-DMF system is not appropriate at least for the preparation of the phase diagram to investigate the validity of the theories. However, the gelation and the gel of rigid rod-like polymers are considered to be quite different from those of flexible polymers in the mechanism and properties and they provide an unsolved, interesting field in polymer physics.

At last we have discussed the cholesteric twisted structure retained in solid films of polypeptides and thermotropic mesophases observed in some copolypeptides. From the latter finding, we suggest that rigid rod-like polymers having dissimilar and/or high density of side chains will manifest thermotropic liquid crystals, in general. Thermotropic liquid-crystalline main-chain polymers are usually realized by alternating rigid segments and flexible segments as known in copolyesters. In the case of thermotropic liquid crystalline copolypeptides, the main chains do not include such flexible segments. Rigid main chains are embedded in the rubbery matrix composed of side chains. This situation will be the origin for the formation of thermotropic liquid crystals in copolypeptides.

Space did not permit us to discuss the electromagnetic orientation of polypeptide liquid crystals. However excellent reviews [76] are now available.

Acknowledgement: The work reviewed in this article was carried out with the cooperation of Drs. S. Sasaki, J. Watanabe and H. Toriumi. We thank them for their enthusiastic cooperation. We also wish to express our sincere thanks to Prof. H. Kimura of Nagoya University for his useful discussion and for permitting us to reproduce figures from Ref. [53].

5 References

1. Elliott, A., Ambrose, E. J.: Disc. Farad. Soc. *9* 246 (1950)
2. Robinson, C.: Trans. Farad. Soc. *52* 571 (1956); Robinson, C., Ward, J. C.: Nature *180* 1183 (1957); Robinson, C., Ward, J. C., Beevers, R. B.: Disc. Farad. Soc. *25* 29 (1958); Robinson, C.: Tetrahedron *13* 219 (1961); Robinson, C.: Mol. Cryst. *1* 467 (1966)
3. Onsager, L.: Ann. N. Y. Acad. Sci. *56* 627 (1949)
4. Flory, P. J.: Proc. Roy. Soc. A *243* 73 (1956)

5. Dimarzio, E. A.: J. Chem. Phys. *35* 658 (1961); Zwanzig, R.: J. Chem. Phys. *39* 1714 (1963); Cotter, M. A., Martire, D. E.: J. Chem. Phys. *52* 1907 (1970); Lasher, G.: J. Chem. Phys. *53* 4141 (1970); Alben, R.: Mol. Cryst. Liq. Cryst. *13* 193 (1971); Straley, J. P.: Mol. Cryst. Liq. Cryst. *22* 333 (1973); Cotter, M. A.: Phys. Rev. A *10* 625 (1974); Cotter, M. A.: J. Chem. Phys. *66* 1098 (1977)
6. Miller, W. G., Wu, L. L., Wee, E. L., Santee, G. L., Rai, J. H., Goebel, K. D.: Prue Appl. Chem. *38* 37 (1974); Miller, W. G., Rai, J. H., Wee, E. L.: Liquid Crystals and Ordered Fluids, Vol. 2. p 243 Plenum, New York, 1974; Wee, E. L., Miller, W. G.: ibid. Vol. 3, p 371, 1978
7. Flory, P. J., Abe, A.: Macromolecules *11* 1119 (1978); Abe, A., Flory, P. J.: ibid. *11* 1122 (1978); Flory, P. J., Frost, R. S.: ibid. *11* 1126 (1978); Frost, R. S., Flory, P. J.: ibid, *11* 1134 (1978); Flory, P. J.: ibid. 1138, 1141 (1978); Flory, P. J., Ronca, G.: Mol. Cryst. Liq. Cryst. *54* 289, 311 (1979); Warner, M. Flory, P. J.: J. Chem. Phys. *73* 6327 (1980); Matheson, R. R., Flory, P. J.: Macromolecules *14* 954 (1981)
8. Flory, P. J., Leonard, W. J.: J. Am. Chem. Soc. *87* 2102 (1965)
9. Rai, J. H., Miller, W. G.: J. Phys. Chem. *76* 1081 (1972)
10. Wee, E. L., Miller, W. G.: J. Phys. Chem. *75* 1446 (1971)
11. Kubo, K., Ogino, K.: Polymer *16* 629 (1975); Kubo, K., Kubota, K., Ogino, K.: ibid. *17* 919 (1976); Kubo, K., Ogino, K.: Mol. Cryst. Liq. Cryst. *53* 207 (1979)
12. Patel, D. L., DuPré, D. B.: J. Chem. Phys. *72* 2515 (1980)
13. Patel, D. L., DuPré, D. B.: J. Polym. Sci. Polym. Phys. Ed. *18* 1599 (1980)
14. Miller, W. G., Kou, L., Tohyama, K., Voltaggio, V.: J. Polym. Sci. Polym. Symposium *65* 91 (1978)
15. Luzzati, V., Cesari, M., Sapch, G., Masson, F., Vincent, J. M.: J. Mol. Biol. *3* 566 (1961)
16. Parry, D. A. D., Elliott, A.: J. Mol. Biol. *25* 1 (1967); Squire, J. M., Elliott, A.: Mol. Cryst. Liq. Cryst. *7* 457 (1960)
17. Watanabe, J., Kishida, H., Uematsu, I.: Polym. Prepr. Japan, *30* (2) 279 (1981)
18. Sasaki, S., Uematsu, I.: Polym. Prepr. Am. Chem. Soc. *20* (1) 106 (1979)
19. McKinnon, A. J., Tobolsky, A. Y.: J. Phys. Chem. *72* 1157 (1968)
20. Fukuzawa, T., Uematsu, I.: Polym. J. *6* 431 (1974)
21. Aritake, T., Tsujita, Y., Uematsu, I.: Polym. J. *7* 21 (1975)
22. Watanabe, T., Tsujita, Y., Uematsu, I.: Polym. J. *7* 181 (1975)
23. Ishikawa, S., Kurita, T.: Biopolymers *2* 381 (1964)
24. Tachibana, T., Kambara, H.: Koll.-Z. Z. Polym. *219* 40 (1967)
25. Sasaki, S., Hikata, M., Shiraki, C., Uematsu, I.: Polym. J. *14* 205 (1982)
26. Marrucci, G., Ciferri, A.: J. Polym. Sci. Polym. Lett. Ed. *15* 643 (1977)
27. Sluckin, T. J.: Macromolecules *14* 1676 (1981)
28. Helfrich, W.: Phys. Rev. Lett. *24* 201 (1970)
29. Wada, A.: J. Polym. Sci. *33* 822 (1960)
30. Toyoshima, Y., Minami, N., Sukigara, M.: Mol. Cryst. Liq. Cryst. *35* 325 (1976)
31. Aikawa, Y., Minami, N., Sukigara, M.: Mol. Cryst. Liq. Cryst. *70* 115 (1981)
32. de Vries, H.: Acta Crystallogr. *4* 219 (1951)
33. Tachibana, T., Oda, E.: Bull. Chem. Soc. Jpn. *46* 2583 (1973)
34. Watanabe, J., Sasaki, S., Uematsu, I.: Polym. J. *9* 337 (1977)
35. Saeva, F. D., Wysocki, J. J.: J. Amer. Chem. Soc. *93* 5928 (1971)
36. Saeva, F. D., Olin, G. R.: J. Amer. Chem. Soc. *95* 7882 (1973); Tsuchihashi, N., Nomori, H., Hatano, M., Mori, S.: Bull, Chem. Soc. Jpn. *48* 29 (1975)
37. Sackmann, E., Voss, J.: Chem. Phys. Lett. *14* 528 (1972)
38. Holzwarth, G., Holzwarth, N. A. W.: J. Opt. Soc. Amer. *63* 324 (1973); Holzwarth, G., Chabey, I., Holzwarth, N. A. W.: J. Chem. Phys. *58* 4816 (1973)
39. Toriumi, H., Yahagi, K., Uematsu, I., Uematsu, Y.: unpublished data
40. Duke, R. W., DuPré, D. B., Samulski, E. T.: J. Chem. Phys. *66* 2748 (1977)
41. Zanker, V.: Z. Physik. Chem. (Frankfurt) *2* 52 (1954); Ballard, R. E., McCaffery, A. J., Mason, S. F.: Biopolymers *4* 97 (1966)
42. Hashimoto, T., Ebisu, S., Kawai, H.: J. Polym. Sci. Polym. Lett. *18* 569 (1980)
43. Picot, C., Stein, R. S.: J. Polym. Sci. A-2 *8* 1491 (1970)
44. Keating, P. N.: Mol. Cryst. Liq. Cryst. *8* 315 (1969)
45. Uematsu, Y., Uematsu, I.: Mesomorphic order in polymers, p156 ACS symposium Series, No 74, American Chemical Society, Washington, D. C., 1978

46. Toriumi, H., Kusumi, Y., Uematsu, I., Uematsu, Y.: Polym. J. *11* 863 (1979)
47. Toriumi, H., Minakuchi, S., Uematsu, I., Uematsu, Y.: J. Polym. Sci. Polym. Phys. Ed. *19* 1167 (1980)
48. Goossens, W. J. A.: Mol. Cryst. Liq. Cryst. *12* 237 (1971)
49. Lin-Liu, Y. R., Yu Ming Shih, Chia-Wei Woo, Tan, H. T.: Phys. Rev. A*14* 445 (1976)
50. Van der Meer, B. W., Vertogen, G., Dekker, A. J., Ympa, J. G. J.: J. Chem. Phys. *65* 3935 (1976)
51. Maier, W., Saupe, A.: Z. Naturforsch. *14*a 882 (1959); ibid. *15*a 287 (1960)
52. Kimura, H., Hoshino, M., Nakano, H.: J. Phys. (France) *40* C3-174 (1979)
53. Kimura, H., Hoshino, M., Nakano, H.: J. Phys. Soc. Jpn. *51* 1584 (1982)
54. Straley, J. P.: Phys. Rev. A*14* 1835 (1976)
55. Czarniecka, K., Samulski, E. T.: Mol. Cryst. Liq. Cryst. *63* 205 (1981)
56. Adams, J. E., Hass, W. E.: Mol. Cryst. Liq. Cryst. *15* 27 (1971)
57. Nagakiri, T., Kodama, H., Kobayashi, K.: Phys. Rev. Lett. *27* 564 (1971); Stegemeyer, H., Finkelmann, H.: Chem. Phys. Lett. *23* 277 (1973); Pochan, J. M., Hinman, D. D.: J. Phys. Chem. *78* 1206 (1974); Kozawaguchi, H., Wada, M.: Jap. J. Appl. Phys. *14* 65 (1975); Bak, C. S., Labes, M. M.: J. Chem. Phys. *62* 3066 (1975); ibid *63* 805 (1975); Hanson, H., Dekker, A. J., van der Woude, F.: J. Chem. Phys. *62* 1941 (1975); van der Meer, B. W., Vertogen, G.: Physics Lett. *74A* 242 (1979)
58. Toriumi, H., Minakuchi, S., Uematsu, Y., Uematsu, I.: Polym. J. *12* 431 (1980)
59. Doty, P., Bradbury, J. H., Holtzer, A. M.: J. Amer. Chem. Soc. *78* 947 (1956); Gupta, A. K., Dufour, C., Marchal, E.: Biopolymers *13* 1293 (1974); Powers, J. C., Peticolas, W. L.: Biopolymers *9* 195 (1970)
60. Watanabe, H.: Nippon Kagaku Zasshi *86* 179 (1965); Kihara, H.: Polym. J. *7* 406 (1977)
61. DuPré, D. B., Duke, R. W., Hines, W. A., Samulski, E. T.: Mol. Cryst. Liq. Cryst. *40* 247 (1977)
62. Patel, D. H., DuPré, D. B.: Mol. Cryst. Liq. Cryst. *53* 323 (1979)
63. Watanabe, H., Yoshioka, K., Wada, A.: Biopolymers *2* 91 (1964)
64. Milstien, J. B., Charney, E.: Biopolymers *9* 991 (1970)
65. Finkelmann, H., Stegemeyer, H.: Ber. Buns...... Phys. Chem. *78* 869 (1974)
66. Bradbury, S. M., Downie, A. R., Elliott, A., Hanby, W. E.: Proc. Roy. Soc., London, *259A* 110 (1960)
67. Samulski, T. V., Samulski, E. T.: J. Chem. Phys. *67* 824 (1977)
68. Toriumi, H., Yahagi, K., Uematsu, I., Uematsu, Y.: unpublished data
69. McKinnon, A. J., Tobolsky, A. V.: J. Phys. Chem. *70* 1453 (1966); Samulski, E. T., Tobolsky, A. J.: Macromolecules *1* 555 (1968); Nature *216* 997 (1967)
70. Fergason, T. L.: Mol. Cryst. *1* 293 (1966)
71. Sasaki, S., Oshima, Y., Watanabe, J., Uematsu, I.: Rep. Prog. Polym. Phys. Japan *21* 553 (1978)
72. Finkelmann, H., Ringsdorf, H., Siol, W., Wendorff, J. H.: Mesomorphic Order in Polymers and Polymerization in Liquid Crystalline Media, p22, ACS Symposium Series 74, Amer. Chem. Soc., Washington, D. C. (1978)
73. Kasuya, S., Sasaki, S., Watanabe, J., Fukuda, Y., Uematsu, I.: Polym. Bull. *7* 241 (1982)
74. So-Lan Tseng., Valente, A., Gray, D. G.: Macromolecules *14* 715 (1980)
75. Shimamura, K., White, J. L., Feelers, J. F.: J. Appl. Polym. Sci. *26* 2165 (1981)
76. Samulski, E. T., Tobolsky, A. V.: Liquid Crystal and Plastic Crystals Vol. I, p175, Gray, G. W., Winsor, P. A. (Ed.) Ellis Horwood (1974); Iizuka, E.: Adv. in Polymer Sci. *20* 80 (1976); Samulski, E. T.: Liquid Crystalline Order in Polymers, Blumstein, A. (Ed.) p167, Academic Press (1978); DuPré, D. B., Samulski, E. T.: Liquid Crystals, Saeva, F. D. (Ed.) p203, Marcel Dekker, (1979); DuPré, D. B.: to be published in "Polymer Liquid Crystals," Ciferri, A., Krigbaum, W. R., Meyer, R. B. (Ed.) p 165, Academic Press

N. A. Platé (guest editor)
Received March 25, 1983

Liquid Crystalline Order in Solutions of Rigid-Chain Polymers

S. P. Papkov, Institute of Chemical Fibers,
U. Kolontzova 5, Mytyschi, Moscow Region USSR

In this review, we discuss cases of formation of polymeric liquid crystalline systems associated with the solutions of so-called rigid-chain polymers.

We do not give a detailed classification of the types of liquid crystals, since the systems under consideration mainly form nematic or in some cases cholesteric phase. The latter phase belongs, in principle, to the same range of liquid crystals as the nematic phase, since there is no phase transition between them (unlike smectic-nematic phase transition).

Introduction 76

1 General Properties of Rigid-Chain Polymers 77

2 Phase Equilibrium in a Rigid-Chain Polymer-Solvent System 81
  2.1 Lyotropic and Thermotropic Liquid Crystals 81
  2.2 Phase Diagrams for a Rigid-Chain Polymer-Solvent System 82
  2.3 Some Properties of Polymeric Systems in the Liquid Crystalline State . . 89

3 Specific Cases of Phase Equilibrium in a Rigid-Chain Polymer-Solvent System 91
  3.1 Phase Equilibrium in the Systems Forming Polymeric Crystallosolvates 91
  3.2 Effect of External Mechanical Fields and the Nature of a Solvent on the Phase Equilibrium 93
  3.3 Kinetics of Phase Transitions in the Systems Involving Rigid-Chain Polymers 95

4 Concluding Remarks 99

5 References 99

Introduction

The liquid crystalline state of polymers has attracted the attention of scientists during last two to three decades. This interest was stimulated, on the one hand, by the works on the peculiar behavior of solutions of poly(γ-benzyl-L-glutamate) in helix-promoting solvents [1-4] and, on the other hand, by a series of works by Soviet and American scientists [5-17] concerning the solutions of aromatic polyamides which contain para-linked phenylene groups in the main macromolecular chain. Poly(γ-benzyl-L-glutamate) is a synthetic analog of natural polypeptides of importance to researchers engaged in biological physics and chemistry. Para-aromatic polyamides are basic polymers for obtaining a new class of high modulus/high strength fibres which are of great practical importance.

It was found that when a certain concentration is attained, the solutions of these polymers acquire thermodynamically stable properties of anisotropy, still retaining relatively high mobility, i.e. they exhibit the properties typical of the liquid crystalline state.

Before these results were published, polymer physicists and chemists mainly investigated only two phase-states, amorphous and crystalline. At the present time, along with these two states, the third phase-state of condensed systems, i.e. the liquid crystalline state, became very important. Here the situation turned out to be the same as in the case of low molar mass liquid crystals. In spite of the fact that historically the low molar mass substances in liquid crystalline state had been known for about a century, the intensive study of their properties began only after they had found an important practical application owing to a sharp change in optical properties of liquid crystals in electromagnetic fields (for visual displays) and as sensitive temperature indicators (in medicine).

Similarly, the ability of polymeric liquid crystals to be easily transformed to highly oriented state and thus to give materials with good mechanical properties stimulated the development of theoretical and applied research. This can be illustrated by the increased number of scientific publications during the last five years.

If we count only the works of real scientific importance, we shall obtain the following figures: 1977-20, 1978-40, 1979-60, 1980-90, and 1981-above 125. A number of monographs, collections of reports at conferences and reviews have also appeared in this field [18-24].

The range of polymers which were found to be able to form liquid crystalline systems has been considerably extended. Poly(γ-benzyl-L-glutamate) and its analogs, as well as para-aromatic polyamides, exhibit this property in solutions, which served the basis for relating them to *lyotropic liquid crystals* (see Sect. 2.1). Subsequently, the classes of polymers were found which exhibited such a transition during a change of temperature (*thermotropic liquid crystals*).

We do not give a detailed classification of the types of liquid crystals, since the systems under consideration mainly form nematic or in some cases cholesteric phase.

The latter phase belongs, in principle, to the same range of liquid crystals as the nematic phase, since there is no phase transition between them (unlike smectic-nematic phase transition).

1 General Properties of Rigid-Chain Polymers

Orientational order appears in the solutions of rigid-chain polymers because a random mutual arrangement of their macromolecules is possible only up to a certain concentration of the solution. To retain a minimal volume (minimal free energy) above a certain critical concentration, asymmetric macromolecules must acquire an ordered mutual arrangement, which corresponds to a transition to the state typical for liquid crystals. In this case the solution becomes anisotropic. The degree of this anisotropy is still less than strict three-dimensional ordering typical of crystalline systems, but at the same time it differs from that of the isotropic state typical of amorphous systems.

The concentration limits for the transition to liquid crystalline state depend on the degree of asymmetry of macromolecules, which is determined as the ratio of their equilibrium length to the diameter. For the most widely used flexible-chain polymers, free intramolecular rotation around valence bonds makes it possible to acquire a random arrangement at any concentration, up to a pure polymer without solvent. However, as free rotation of the chain units around single bonds is being gradually restricted, the macromolecules become more and more asymmetric. In the limit (which is, however, never attained) this should result in a conformation for which the total length of a macromolecule is equal to the length of a completely extended chain.

There can be the following causes of a sharp asymmetry of macromolecules:

1) the restrictions imposed onto a free rotation of units, which may appear owing to the inclusion into a backbone chain of cyclic units (e.g., phenyl units in aromatic polyamides and polyesters, or glucopyranosic units in cellulose and its derivatives)
2) cyclization through intramolecular hydrogen bonds (e.g., in poly(γ-benzyl-L-glutamate) and other polypeptides)
3) quasi-conjugation and coplanarity of amide groups (in poly(alkylisocyanates))
4) the formation of so-called ladder polymers.

The conformation of macromolecules in linear polymers can be characterized by the vector h which connects the chain ends (end-to-end vector). For rigid-chain polymers the conformation of macromolecules is usually described by the persistence length a (worm-like chain model) which is related to h as:

$$h^2/(2aL = 1 - (1 - e^{L/a})/(L/a)$$

where L is the contour length of a macromolecule.

The rigidity of a chain is characterized by the statistical segment A introduced by Kuhn [25]. For long chains the Kuhn segment is equal to the doubled persistence length ($A = 2a$).

In order to give a general idea about the appearance of liquid crystalline state and about the properties of the systems containing these polymers, some examples of linear rigid-chain polymers follow:

Poly(γ-benzyl-L-glutamate) (PBLG), whose structure is given by the formula

$$\left[CO - - - - C - - - - NH \right]_n$$

with H above C and CH$_2$·CH$_2$COOCH$_2$·C$_6$H$_5$ below

may serve as an example of a polymer with a high axial ratio (asymmetry) of the macro-molecules.

This polymer may in principle be referred to the class of flexible chain polymers, but owing to intramolecular hydrogen bonds appearing between the NH-group and the fourth CO-group along the chain in those solvents which stabilize (and do not break) this hydrogen bond, the α-helical conformation is formed, which imparts to the entire macromolecule the state of a rigid rod. In addition to its importance as a peculiar model of natural polypeptides, this polymer is the object of investigation of the liquid crystalline state, since owing to the helix-coil transition, the change in the type of a solvent makes it possible to demonstrate directly the role of the asymmetry of particles in the appearance of liquid crystalline order.

The solution of *para-aromatic polyamides* occupy a special place among the systems forming liquid crystalline order. Two typical representatives of this class of polymers turn out to be best studied owing to their practical importance. These are *poly-(p-benzamide)* (PBA)

$$\left[CO - \bigcirc - NH \right]_n$$

and *poly(p-phenylene terephthalamide)* (PPTA)

$$\left[CO - \bigcirc - CONH - \bigcirc - NH \right]_n$$

These two polymers are characterized by a stable trans-configuration of the amide bond and by a high barrier of the rotation around the aryl carbon bond. The directions of rotational axes in all the chain units almost coincide. As a result, according to Tsvetkov [26], the entire macromolecule acquires the shape of a crankshaft (Fig. 1). In the case of PBA, the retardation of rotation of the units may be also caused by a transfer of conjugation in the carbonyl-nitrogen system via a phenylene ring.

Theoretical calculations of dimensions of the chains of para-aromatic polyamides were made by Birshtein [27]. Persistence length was estimated within the limits of

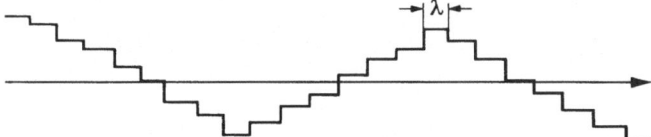

Fig. 1. "Crankshaft" conformation of rigid-chain macromolecules (according to Tsvetkov [26]). λ — a structural unit

400–500 Å. Recently, Erman, Flory, and Hummel [28] carried out a detailed analysis of end-to-end vectors as a function of the chain length for p-phenylene polyamides and polyesters. The required geometrical parameters were obtained from the results for model compounds. Configurational averaging was performed on the basis of torsional potentials obtained by the same authors [29]. The value of persistence length depends on the difference between the angles of amide bond of CO and N. Taking this difference equal to 10°, the authors estimated the persistence length as having 410 Å. It was pointed out that this result should be regarded as the upper limit. With the account of torsional fluctuation effects and bond angle bending, mentioned by Birshtein [27], this value may be lowered by 25–40 %.

The dimensions of chain macromolecules have been experimentally estimated by various methods. Unfortunately, the method of sedimentation by ultracentrifuge, which is usually employed for the study of macromolecular conformation, is insufficiently sensitive for rigid-chain polymers because of weak dependence of the sedimentation coefficient [S] on the molecular mass. M. Prozorova et al. [30] have found for PBA in dimethylacetamide (+LiCl) that upon the change in M from $13.8 \cdot 10^3$ to $52.5 \cdot 10^3$, the value of [S] changes from 0.5 to 0.65 Svedberg units. The Kuhn segment was estimated in this work as having 380–390 Å (persistence length $a = 190$–195 Å).

Using the method of flow birefringence, Tsvetkov [26] estimated the Kuhn segment for PBA in sulphuric acid as beving 2100 Å ($a = 1050$ Å), and for PPTA, as 1300 Å ($a = 650$ Å). In another work [31] he points the Kuhn segment for PBA being equal to 1960 Å ($a = 980$ Å) and in a later work [32], 1000 Å ($a = 500$ Å) for PBA and 600 Å ($a = 300$ Å) for PPTA.

Similar investigations were carried out by Arpin and Strazielle [33]. They assessed the persistence length for PBA and PPTA as of the order of 400 and 200 Å, respectively.

Lower values were obtained when the presistence length was estimated by the beginning of a sharp kink on the η_{sp}/c vs. c plot (c is the concentration). This kink indicates the overlapping of the rotation spheres of asymmetric particles. The persistence length for PBA was found to be equal to 180–240 Å, and for PPTA, 150–180 Å [34].

In spite of the considerable difference in experimental and theoretical results obtained by different authors for the dimensions of macromolecules of aromatic polyamides, these results indicate a sharp asymmetry of the macromolecules. It should be pointed out, for the sake of comparison, that for flexible chain polymers the Kuhn segment amounts to only 20–50 Å. For example, for an aromatic polyamide which does not exhibit coaxial rotation of the bonds, poly(m-phenylene isophtalamide), the Kuhn segment is equal to 50 Å, i.e. it is 10–15 times as small as that for its para-analog.

Another indication of a rigid-chain nature of polymers under consideration is the value of the exponent a of the molecular mass M in the Mark-Houwink Equation for the intrinsic viscosity $[\eta]$:

$$[\eta] = KM^a$$

According to theoretical calculations, the exponent a varies from 0.5 for flexible chain polymers to 1.8 for rigid-chain polymers. This exponent, calculated in [9] on the assumption of the maximum rigidity for PBA, turned out to be equal to 1.85. More rigorous experiments and calculations have shown that this value, as it should be expected, is below the maximum value. The value of this exponent depends on the molecular mass. For rigid-chain polymers with a high molecular mass the macromolecular conformation approximates the Gaussian coil and a diminishes to 1.0–1.1, while for the macromolecules with a low molecular mass, whose conformation is close to linear, a approaches 1.7–1.75.

Naturally, the macromolecular conformation and hence the intrinsic viscosity also depend on the type of a solvent.

To conclude this paragraph, we shall mention one more feature of rigid-chain polymers. One of the reasons why the appearance of liquid crystalline order in the solutions of these polymers was established only three decades ago is their *low solubility* in the solvents normally used for polymers. Most polymeric materials used in practice are processed through the melt, i.e. using the most technologically and economically advantageous method. The melting point of flexible chain polymers is relatively low and as a rule does not reach the region where intensive thermal degradation of a polymer takes place. On the contrary, rigid-chain polymers have very high melting points and may be processed only in solution. But even the solubility of rigid-chain polymers is confined to a relatively narrow range of solvents, since the entropy term of the change in the free energy makes very small contribution. Dissolution of such polymers is predominantly attained owing to intensive interaction of a polymer with solvent molecules (the enthalpy term of the free energy).

Relatively high solubility of PBLG in many solvents can be explained by the interaction of flexible side groups with solvents. On the other hand, for aromatic polyamides with *p*-structure, e.g. PPTA, the solubility is observed only for a very narrow range of solvents, mostly for concentrated acids (sulphuric, phosphoric, chlorosulphonic, hydrofluoric, and other acids). In this case the interaction of acid molecules with amide groups of a polymer reaches the energy of chemical reactions.

Limited solubility creates considerable difficulties both for technological processing of rigid-chain polymers into articles and for scientific investigations. In particular, the formation of additive compounds of acids with amide groups in polyamides results in some cases in the appearance of crystallosolvates precipitating in the form of crystalline spherulites at ordinary temperatures [35]. We shall discuss this in greater detail when considering the factors which complicate the phase diagrams for rigid-chain polymer-solvent systems and which mask the transitions into liquid crystalline state (see Sect. 3.1).

We confined ourselves only to poly(γ-benzyl-L-glutamate) and *p*-substituted aromatic polyamides, as most of experimental studies on phase equilibrium with the formation of liquid crystalline phase are based on the analysis of the behavior of these polymers which can be regarded as typical (model). For these polymers we can compare theoretical results with experiment. As to other polymers, which have lower rigidity of the chains and go over to the liquid crystalline state at extremely high concentrations of a polymer in solution, only few experimental data have been published so far (some of them are considered in other sections of this review).

2 Phase Equilibrium in a Rigid-Chain Polymer-Solvent System

2.1 Lyotropic and Thermotropic Liquid Crystals

In principle, *lyotropic* liquid crystalline systems are such systems in which liquid crystalline ordering appears only in the presence of a solvent, the solvent being selective and the dissolved substance having amphiphylic properties. Soaps, lipids, and some dyes in an aqueous medium may serve as examples of such systems.

For rigid-chain polymers, where the main units and side groups adjoining them are arranged, as a rule, uniformly along the entire chains, the liquid crystalline state is not associated with the ordering owing to the selective action of a solvent and may appear generally without a solvent. However, high melting points and high viscosities of melts prevent from direct transition into liquid crystalline state without diluting a polymer.

In this connection let us consider a fragment of a schematic phase diagram in the region of high concentrations of a polymer capable of forming the liquid crystalline phase (Fig. 2). In a crystalline polymer containing no solvent (100 % polymer, v_2^0), the transition from the crystalline state (c) to the liquid crystalline state (lc) must take place at the temperature $T_{c \to lc}^0$ and further transition into isotropic state (i), at the temperature $T_{lc \to i}^0$. Such transitions are called *thermotropic*, and the system formed at $T_{c \to lc}^0$ is called the thermotropic liquid crystal. The transition to the liquid crystalline state can also occur by adding to a polymer a solvent at a temperature below $T_{c \to lc}^0$.

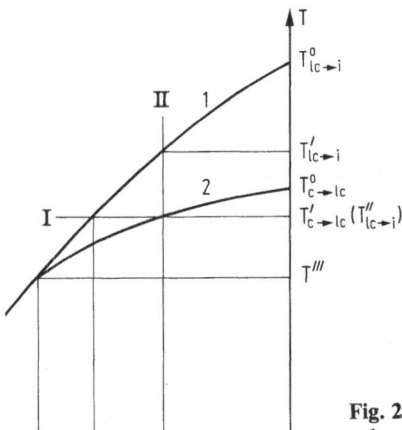

Fig. 2. A portion of the phase diagram for mesogenic polymer-solvent system

For example, if the polymer is diluted by introducing a solvent to the concentration v_2^0, the system gets into the liquid crystalline state at the temperature $T'_{c \to lc}$ (route I). Upon further diluting the polymer at the same temperature the system goes over to the isotropic state with the polymer concentration v_2''. Similar transitions, however, can take place by heating the system at a given polymer concentration. For example, when the temperature is increased at the concentration v_2' (route II), the successive transitions c → lc (at $T_{c \to lc}$) and lc → i (at $T'_{lc \to i}$) take place. Just as in the case of a pure polymer, such transitions should be regarded as thermotropic.

If a polymer is incapable of crystallizing or if it has retained non-crystalline form for kinetic reasons, it still remains in the liquid crystalline state in the entire region of temperatures and concentrations below curve 1 and not only between curves 1 and 2. The point $v_2''' T'''$ corresponds to the limiting concentration and limiting temperature for the existence of liquid crystalline state of the polymer.

Thus, the terms "thermotropic" and "lyotropic" transitions are correct, but there are no grounds behind the division of liquid crystalline polymers into lyotropic and thermotropic (with the exception of block copolymers which form, in selective solvents, systems that can conventionally be referred to as liquid crystals [36]).

A peculiar property of rigid-chain polymers is that their melting point lies above the temperature of intensive thermal degradation. For this reason, the liquid crystalline state is realized only in diluted systems when the melting point of a polymer is lower then the experimental temperature. For non-crystallizing polymers, the liquid crystalline state is possible in a wide range of concentrations, including pure polymer, at temperatures below the boiling point of a solvent and below the temperature of intensive thermal decomposition of a polymer.

However, at a low content of a solvent in the system, the viscosity becomes so high that it is difficult to distinguish between the true (thermodynamically stable) anisotropic state of the system and the artificially created (thermodynamically unstable) oriented state which appears upon mechanical deformation of such systems and which very slowly relaxed owing to a high viscosity of the system.

2.2 Phase Diagrams for a Rigid-Chain Polymer-Solvent System

As applied to colloid systems, the problem of the ordering of rigid rods in a liquid medium was considered for the first time by Onsager [37]. He showed, by minimizing the expression for the free energy of such systems, that when the volume fraction of rodlike particles in the system exceeds a certain value (v_2^*), two minima appear on the free energy vs. composition curve, and the system separates into two phases. In one of the phases the particles are arranged randomly with respect to each other (isotropic phase), while in the other they are ordered (anisotropic phase). This ordering has a nematic nature, i.e. in the system there exists a long-range orientation order and there is no translational order. As the particle concentration increases, the second limit v_2^{**} is attained, at which the isotropic phase vanishes and the system becomes completely anisotropic. The composition of equilibrium phases v_2^* and v_2^{**} varies with a change in the asymmetry x (the ratio of the length to the diameter) of the particles as follows:

$$v_2^* = 3.34/x, \qquad v_2^{**} = 4.49/x$$

Later on [38] these values were defined more exactly.

Flory [39] presented a more detailed analysis of the molecular theory of the liquid crystalline state of polymeric systems. This theory is discussed in the first chapter of this book, so we here shall only describe the principle of calculating the equilibrium phase composition without going into details. Note that Flory's theory embraces a wide range of concentrations, while Onsager's analysis is valid only for low concentrations, since it is based on the second virial approximation.

On the basis of the lattice model Flory calculated the entropy term of the change in the free energy for mixing molecules of solvent and macromolecules of polymer with the degree of asymmetry (axial ratio) x. In these calculations, a parameter y was introduced, which is a measure of disorientation of macromolecules. This parameter varies from unity for the ideal order to x for a complete disorder. Taking into account the van Laar heat of mixing as a function of Huggins-Flory interaction parameter χ_1, the general expression for the change in the free energy ΔG_M, can be obtained. For a given value of x the change in the free energy of mixing is a function of the mixture composition and of the parameter y. For a completely disordered state, i.e. for $y = x$, the equation transforms into the well-known Flory-Huggins equation for flexible chain polymers. For $y \neq x$, an analysis of the equation for ΔG_M shows that above a certain critical concentration, the anisotropic state becomes stable. The critical volume concentration of polymer v_2^* varies with the asymmetry of macromolecules.

Above the critical concentration v_2^* the solution becomes metastable and separates into two phases — isotropic and anisotropic. The condition of thermodynamic equilibrium of the two phases corresponds to the equality of the chemical potentials of each of the components in each of the coexisting phases. The concentration corresponding to a complete transition to the anisotropic state, v_2^{**}, is 1.56 times as high as v_2^* (see also [40]).

Figure 3 shows the phase diagram calculated by Flory in v_2 vs. χ_1 coordinates for $x = 100$. It can be seen from the diagram that there are three regions of phase state of the system. In region I (corresponding to a low polymer concentrations) the system is monophase and the solution is isotropic. The macromolecules are arranged random-

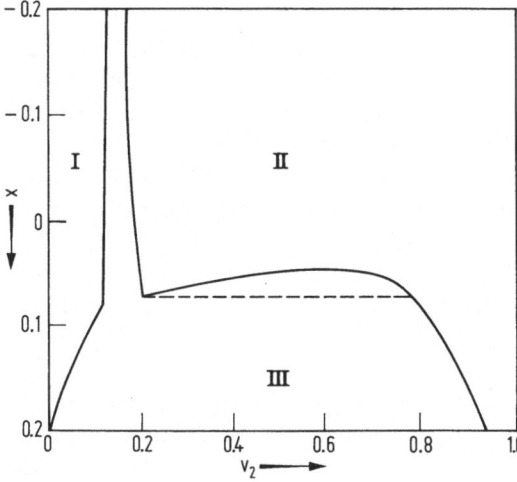

Fig. 3. Phase diagram for $x = 100$ (according to Flory [39])

ly with respect to each other. In region II the solution is also monophase, but because of the lack of free volume, rigid macromolecules are forced to take mutually ordered state. The solution exhibits anisotropic properties (including optical properties). Region III is heterophase and consists of two phases. This region separates isotropic and anisotropic monophase solutions. For χ_1 below ~ 0.07, the intermediate hetero-phase region is very narrow ($v_2^{**}/v_2^* \approx 1.5$). But above $\chi_1 = 0.07$, the system separates into the phases which differ widely in concentration. The concentration of polymer in phase I becomes very low, while in phase II it is very high.

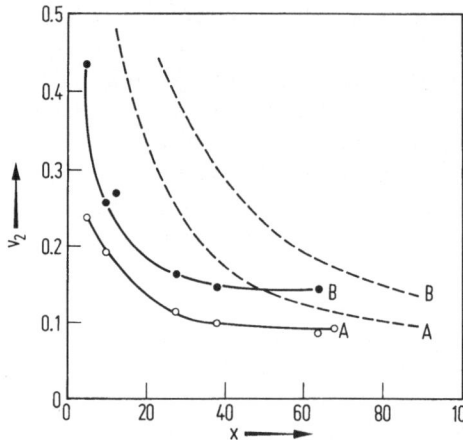

Fig. 4. The composition of coexisting phases (A and B are isotropic and anisotropic phase, respectively) for a poly(γ-benzyl-L-glutamate)-dioxane system (according to [41]). Solid curves — experiment, dashed curves — theory

It is important to note that the theoretical results obtained by Flory [39] about the dependence of the critical concentrations v_2^* and v_2^{**} on x are in good agreement with experimental data. It is sufficient to remember, as an example, the results obtained by Flory [41] for PBLG solutions in dioxane (Fig. 4). The discrepancy between the experimental results (solid curves) and theoretical calculations (dashed curves) looks quite natural on the account of a number of assumptions made when deriving the equation for the free energy on the basis of the lattice model (see also [42]).

For para-aromatic polyamides, the experimental results have also been obtained that confirm the existence of phase transitions from isotropic solutions to anisotropic ones via the region of coexistence of the two phases. This is illustrated by Fig. 5 [43] (see also [11-15]).

Further theoretical studies by Flory [44-50] concerned the clarification of phase transitions for the systems with a definite distribution of macromolecules over the length and also the analysis of equilibrium for a model system composed of macromolecules in which rigid blocks are separated by flexible units. An example of such systems are copolyesters exhibiting the thermotropic transition into the liquid crystalline state in the absence of a solvent [51, 52].

This system of rigid blocks with flexible spacers may serve as a model of polymers with a limited flexibility. In his early work Flory [53] considered the behavior of semiflexible chain polymers by introducing the flexibility parameter f which represents the fraction of bonds which are not in a collinear position in the

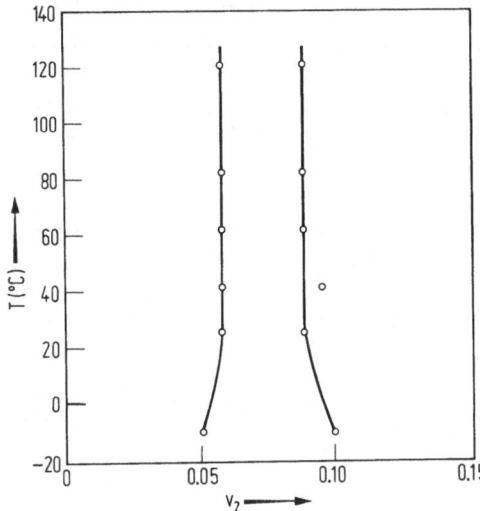

Fig. 5. Phase equilibrium in a PBA-dimethyl-acetamide (+LiCl) system (according to [43])

chain (inflections). Using this parameter, Flory calculated the free energy of mixing a polymer with a limited flexibility of chains and the solvent molecules and established the critical value of the parameter f ($f_c = 0.63$), which determines the stability of disordered and ordered states (see also [54]).

Even a small deviation of f from zero results in a considerable broadening of the heterophase region in the phase diagram. This calculation was made by Miller et al. [55] for $f = 0.1$. It turned out that for $x = 100$, the concentration v_2^* shifts from ~ 0.08 to ~ 0.165 (volume fractions). Werbowyi and Gray [56] extended these calculations to $f = 0.2$, 0.4, and 0.5. As is shown in Fig. 6, the values of v_2^* in this case are equal to 0.4, 0.65, and 0.8 respectively.

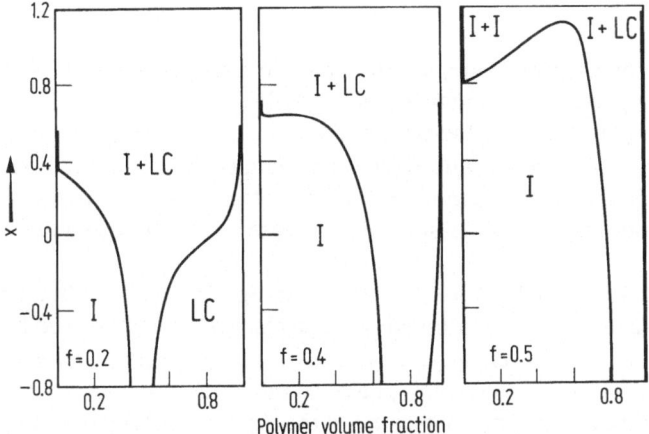

Fig. 6. The effect of the variation of the flexibility parameter for a semirigid-chain polymer (at $x = 100$) on the composition of coexisting phases (according to [56])

Returning to the analysis of the typical phase diagram for a rigid-chain polymer-solvent system shown in Fig. 3, we note that for $f = 0$, according to theoretical calculations, the heterophase region at athermal conditions ($\chi_1 = 0$) for x = 100 lies within the limits from 0.079 to 0.116 and slightly changes with χ_1 up to $\chi_1 \approx 0.07$. Above this value of χ_{19} a broad heterophase region appears. The results obtained by Nakajama et al. [57] on experimental construction of the phase diagram for solution of PBLG (x = 150 and 350) in dimethylformamide may serve as an experimental corroboration of such a transition from a narrow to a wide region of coexistence of two equilibrium phases. Increasing χ_1 by introducing methanol into the system, they demonstrated the transition from a narrow to a wide heterophase region for the methanol content of 0.10 to 0.12 volume fractions (Fig. 7).

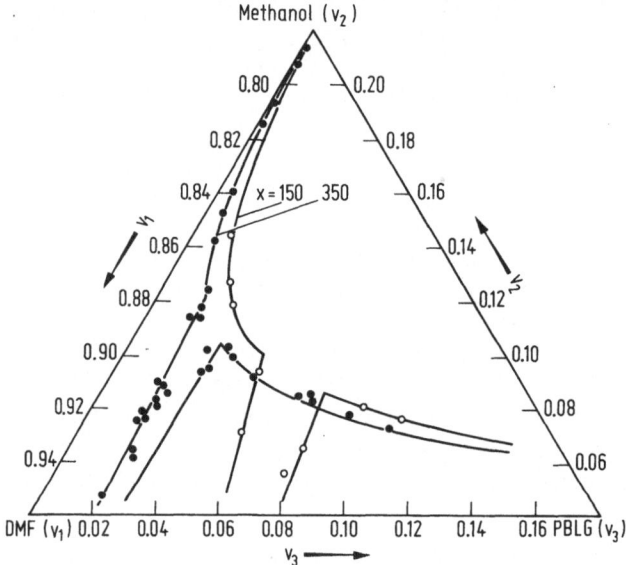

Fig. 7. Phase diagram for a PBLG-dimethylformamide-methanol system (according to [57]): $\bigcirc - x = 150$; $\bullet - x = 350$

The transition to a wide heterogeneous region may be also attained by decreasing the temperature. Such an approach was demonstrated by Miller et al. [42] for a PBLG-dimethylformamide system (Fig. 8). As to the systems containing para-aromatic polyamides, this transition to a wide region with decreasing temperature has not been clearly observed because of the crystallization of the solvent and in some cases of cocrystallization of a polymer together with a solvent. This case will be discussed more in detail in Sect. 3.1.

The elevation of temperature usually leads to a slight change in χ_1. According to Flory, a narrow heterophase region in this case also varies slightly (v_2^* and v_2^{**} just come a little closer one to another). However, in real systems it is impossible to observe experimentally the change in the equilibrium phase composition because of the attainment of the boiling points of solvents or because of the beginning of rapid chemical transformation of the polymers (e.g., oxidative degradation of polyamides in sulphuric acid).

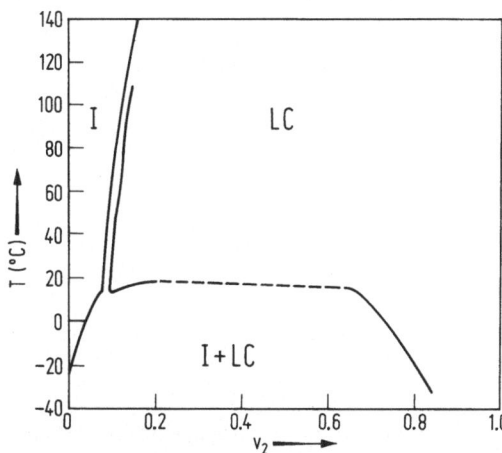

Fig. 8. Phase diagram for a PBLG (M = 310000)-dimethylformamide system. Dashed line indicates the region where experimental data are not sufficiently clear (according to [42])

Miller et al. [42] observed for PBLG-dimethylformamide systems a more noticeable change in the concentrations v_2^* and v_2^{**} with increasing temperature than follows from Flory's analysis. It can be seen from Fig. 8 that there is a shift of equilibrium concentrations towards a higher polymer content. A similar shift is also observed for a PPTA-sulphuric acid system in the region of elevated temperatures [18].

Without giving a detailed analysis of possible causes of this discrepancy between the experimental curves for $v_2^* - v_2^{**}$ and theoretical dependence predicted by Flory, we only note that an increase in the temperature results in an increase of the free volume of the system. Hence the limits at which disordered mutual arrangement of macromolecules becomes impossible should also be shifted. In pure polymers containing no solvent a transition from the liquid crystalline state to the isotropic melt is also connected with the attainment of such a free volume for which the molecules do not need to be mutually oriented. In this connection let us consider a hypothetical phase diagram in the composition range from $v_2 = 0$ to $v_2 = 1$ with the temperature increasing up to $T_{lc \to i}$ (which cannot be attained in practice). In this hypothetical case the hetereophase region at the point $T_{lc \to i}$ should disappear as we attain $v_2 = 1$. Then the phase diagram will have the form schematically shown in Fig. 9 (see [58]). Thus, phase compositions v_2^* and v_2^{**} in principle must have a tendency to simultaneously increase with the temperature. However, as we have already pointed out, this hypothetical case cannot be realized because of the limitations imposed on thermal stability of rigid-chain polymers and because of the volatility of solvents at normal pressures.

When analyzing the deviations of phase equilibrium for real polymeric systems forming a mesophase from the theoretically calculated phase diagram, we must pay attention to one more circumstance, namely, polydisperse nature of real polymers. In all the cases the transition from an isotropic solution to anisotropic one is a result of superposition of equilibria typical of individual fractions of a polymer, which differ in molecular mass. As was shown by Flory [45−48], this must result in a broadening of the heterophase region $v_2^* - v_2^{**}$. Indeed, as was shown in early works by Kwolek et al. [12,14], the equilibrium phases v_2^* and v_2^{**} differ in average molecular masses. High-molecular fractions are accumulated in the phase v_2^{**} just as in the case of liquid (amorphous) equilibrium in a flexible chain polymer-solvent system [59].

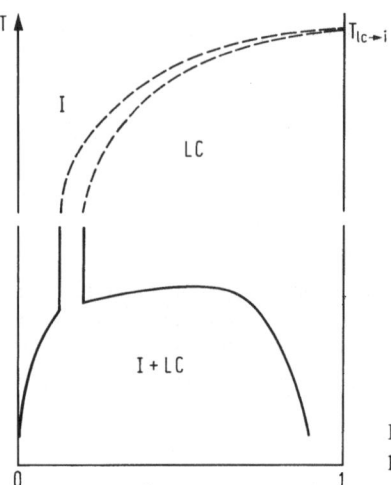

Fig. 9. A hypothetical phase diagrams for a rigid-chain polymer-solvent system in a wide temperature range [58]

As has been mentioned above, the methods of statistical mechanics used by Flory for the theoretical calculation of the change in the free energy of mixing rigid rods and a solvent are based on a number of assumptions. Nevertheless, as can be seen from a comparison of theoretical results with experimental data, the main conclusions are qualitatively in coincidence. Critical remarks made in publications after the works of Flory and Onsager mainly dealt with some details [60-62].

Indeed, Flory's theory in its initial form considered phase separation on the basis of the determining role of the entropy term of the free energy with no account of the forces acting among the particles. In later works [40,63], however, orientation-dependent interactions of particles were taken into account, which increased the probability of the liquid crystalline transition.

Naturally, further studies aiming at the refining of the quantitative relations for phase equilibrium in a rigid-chain polymer-solvent system are very important. It should be noted, however, that since Flory and other researchers have established and experimentally verified the general form of the phase diagram for polymeric systems involving the formation of liquid crystalline phases, it is very important to carry out the geometrical (topological) analysis of phase equilibria, which is widely used for multicomponent systems in condensed state. In this connection we refer to Konstamm's remark made in "Lehrbuch der Thermostatik" by Van der Waals [64]. He wrote that what physicist and chemist need is not at all the quantitative dependence for a specific case. It is sufficient to establish the general types and then to find out whether the qualitative differences in these types coincide with the cases obtained experimentally. For this the method of analysis is required, which mathematically should rather be called the geometry of positions (topology, analysis situs) and not the theory of functions, if by analysis situs we mean, after Klein, the purely position relations which do not depend on metrical quantities.

From this point of view Flory's works give a reliable basis for a general analysis of phase equilibrium in polymeric systems involving the formation of liquid crystalline phases. In particular, this applies to the cases when these equilibria are complicated by additional factors. Some examples of these complications will be considered in Chapter 3.

2.3 Some Properties of Polymeric Systems in the Liquid Crystalline State

The establishment of the fact that a polymeric system has been transformed to the liquid crystalline state does not differ in principle from that for low-molar mass liquid crystals. The main characteristic of the liquid crystalline state is the appearance of anisotropic properties of the system, and above all optical anisotropy, i.e. the ability to transmit light between crossed polaroids (birefringence). In this respect liquid crystals are similar to three-dimensional crystals, though they differ from the latter in that they have no long-range translational order and retain only orientational order. But this order is not strict. There exists a certain distribution over the angles with respect to the principal direction (director). The degree of ordering is characterized by the orientational parameter s:

$$s = \frac{1}{2}(3\langle\cos^2\theta\rangle - 1)$$

which may vary from 0.4 to 0.9.

It has already been mentioned above, however, that we should distinguish between the true anisotropy and induced, unstable, anisotropy especially for superconcentrated solutions and a solid polymer subjected to mechanical effects.

Special attention should be paid to *rheological behavior* of liquid crystalline polymeric systems, which although being similar to that of low-molecular liquid crystals, still has some peculiarities. Without going into a detailed discussion of rheological properties, which are described in separate articles and reviews [11, 65, 66], we shall mention some of their features.

A transition into the liquid crystalline state with increasing concentration of a rigid-chain polymer in a solution is characterized by a sharp peak of viscosity, behind which the viscosity abruptly drops. This is shown in Fig. 10 [11], where the change in viscosity is given in η/η^* vs. c/c^* coordinates. Although for some polymers there is no exact coincidence nevertheless the peaks on the viscosity vs. concentration curve closely corresponds to the transition from isotropic solution to the region of coexistence of isotropic and anisotropic phases. With a further increase in the concentration and a transition into the region of only anisotropic phase, the viscosity abruptly increases again. It is not excluded that the minimum viscosity corresponds to the phase inversion region in the heterogeneous system.

One more distinguishing feature of liquid crystalline polymeric systems is a peculiar dependence of the viscosity on the shear stress (τ) when the latter is small. Kulichikhin et al. [67,68,11] were the first to note that in the range of small shear stresses the viscosity strongly depends on the stress. This indicates the existence of the yield point (Fig. 11). This effect was later confirmed by Onogi and Asada [69]. With a further increase in the shear stress the viscosity remains constant in a certain interval of τ, and then a new decrease in viscosity is observed.

These peculiarities in the behavior of polymeric liquid crystals indicate a specific nature of such systems and call for further detailed studies.

We shall mention here another property of liquid crystalline polymeric systems. As in the case of low-molar mass liquid crystals, when electric and magnetic fields are applied, *liquid crstalline domains get oriented along the direction of the field*. Rearrangement of a polymer structure under the effect of a magnetic field was demonstrated in [15, 70, 71] for a PBA-dimethylacetamide system. However, the processes of

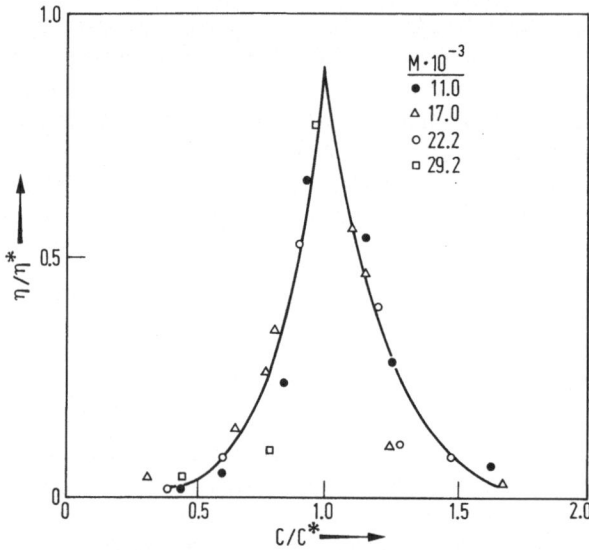

Fig. 10. The concentration dependence of viscosity for a PBA solution in dimethylacetamide (+LiCl): C* is the critical concentration of the transition into the liquid crystalline state, η^* is the maximum viscosity at the point of the liquid crystalline transitions (according to [11])

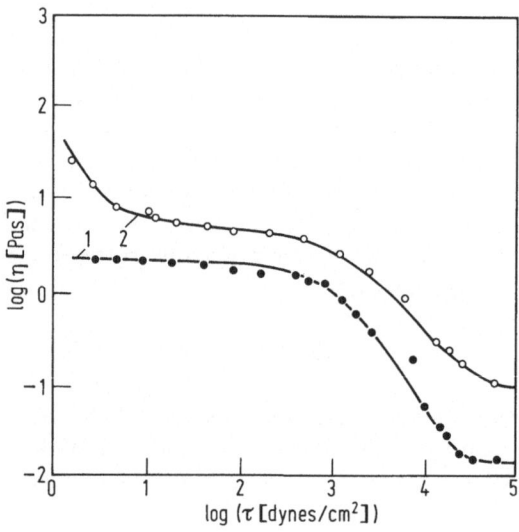

Fig. 11. The dependence of the viscosity ($\log \eta$) on the shear stress (τ) for a solution of PBA in dimethylacetamide (+LiCl). Curve 1 corresponds to isotropic solution ($v_2 = 3\%$); curve 2, to anisotropic solution ($v_2 = 7\%$ at $v_2^* = 5\%$)

orientation and reorientation in a magnetic field for the systems under consideration occur much more slowly than for low-molar mass liquid crystals because of high viscosity of the former systems. For this reason, at the present time, it is hardly possible to use this effect in the same technical devices where low-molecular liquid crystals are employed. This explains the absence so far of further detailed studies of the effect of magnetic fields on polymeric liquid crystals.

Aikawa et al. [72] considered the effect of electric field on the phase transition in solutions of rigid-chain polymers for a PBLG solution in dioxane. Theoretical calculations have shown that the application of an electric field must shift the values of v_2^* towards lower concentrations. This conclusion was confirmed in experiments. According to the results obtained by Patel et al [73], the application of electric fields also causes a shift in the temperature of the liquid crystalline transitions.

Principal properties of linear rigid-chain liquid crystal systems

1. Transition into the liquid-crystalline state by reaching some critical concentration of polymers
2. Optical anisotropy of solution above this critical concentration
3. Nematic or cholesteric type of the mesophases (by absence of the smectic one)
4. Abrupt dropping of the viscosity by transition into the liquid-crystalline state
5. Existence of the yield point by small shear stresses
6. Ability of orientation and reorientation of macromolecules along the direction of the mechanical, magnetic and electric field
7. Slowness of the relaxation processes in comparison with the low-molecular liquid-crystals
8. Ability of forming fibres with the thermodynamic stable orientation of macromolecules along the fiber axis (and consequently with high strength and high modulus)

3 Specific Cases of Phase Equilibrium in a Rigid-Chain Polymer-Solvent System

3.1 Phase Equilibrium in the Systems Forming Polymeric Crystallosolvates

Only a limited number of solvents can be used for rigid-chain polymers because of small changes in the entropy term of the free energy of mixing a polymer with a solvent. For this reason, the solvents used predominantly are those which interact with active groups in the polymer chains very intensively.

The formation of crystallosolvates is typical for a number of polymeric systems in which a polymer actively interacts with a solvent (for example, polyamides + H_2SO_4 [74,75]). A review paper in this field was recently published [76].

The systems involving rigid-chain polymers and forming both the liquid-crystalline and the crystallosolvate phase are characterized by a complex phase equilibrium. The general principles of constructing phase diagrams (topological analysis) allow us to assume [59,77] that the sequence of phase transformations in such systems has the form schematically shown in Fig. 12. This schematic diagram illustrates incongruent melting of crystallosolvate, i.e. melting accompanied by decomposition of an additive compound.

In the region lying below T_1 and up to the polymer content v_{cs} in the system which corresponds to the crystallosolvate composition, there is phase equilibrium between the isotropic solution and crystallosolvate (region I + CS). Curve 1 describes the transition from the heterogeneous region I + CS to the region of isotropic mono-phase solution up to the polymer content v_2'. In the interval $v_2' - v_2''$ the fusion temperature of crystallosolvate remains constant (T_1) (curve 1'). Above this curve there is the region of coexistence of the two phases-isotropic and liquid crystalline (I + LC). The position of this region is determined by curves 2 and 2'. As the polymer content in the system exceeds v_2'', the fusion temperature of crystallosolvate varies along the curve 3 which separates the region of monophase liquid crystal, LC, from the hetero-phase region LC + CS.

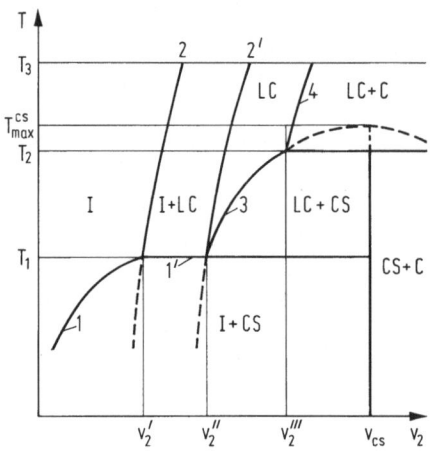

Fig. 12. Schematic phase diagram for a rigid-chain polymer-solvent system involving the forma-tion of the liquid crystalline phase and crystallo-solvate (according to [77])

For congruent melting of crystallosolvate (i.e. without its decomposition into components) the maximal fusion temperature is attained (T_{max}^{cs}). In Fig. 12 this is illustrated by the dashed line which is a continuation of curve 3. If the energy of for-mation of additive compound is not very high, the decomposition of crystallosolvate (incongruent melting) occurs before the temperature T_{max}^{cs} is reached. This is the case illustrated by Fig. 12, where incongruent melting begins at the temperature T_2 and at polymer content v_2'', i.e. before the concentration v_{cs} is reached. Curve 4 separates the monophase region LC from the heterophase region LC + C. The separation of crystalline polymer C above T_2 and concentration v_2'' is associated with the above-mentioned decomposition of a polymer-solvent additive compound. It is not excluded that in some cases crystallosolvates may be formed with other molar ratios n:m.

In the heterophase region LC + C the system goes over to the solid state typical of crystalline substances. Usually this solidification of the system is treated as a limited solubility of a polymer. For para-aromatic polyamides the solid state lying beyond curve 4 may already be attained at polymer concentration of 20–25% (by weight). Along with this "solubility limit" of a polymer, there also exist a limiting temperature above which either boiling of a solvent occurs, or (which is a more frequent case) rapid thermal degradation of the polymer itself takes place (T_3).

The phase diagram shown in Fig. 12 is just approximate and requires experimental verification. Note that experimental investigations in this case are very complicated because of low rates of certain phase transitions, which results in the formation of nonequilibrium systems. Some examples illustrating the role of kinetics of the processes in the establishment of phase equilibrium will be considered in Sect. 3.3. This applies, in particular, to the transformation occurring in the region lying between the continuations of curves 2 and 2' (dashed line) below the temperature T_1. In this case the phases I and CS should exist in the equilibrium state, while in nonequilibrium state the phase LC is present, which vanishes only after a long time by transforming into the phases I and CS.

Undoubtedly, the studies of polymeric crystallosolvates in general and in combination with liquid crystalline transition in particular are very important from the scientific and practical viewpoint.

3.2 Effect of External Mechanical Fields and the Nature of the Solvent on Phase Equilibrium

Phase equilibrium in a rigid-chain polymer-solvent system may be considerably affected by various factors. In particular, this applies to
(1) the application of an external mechanical field under effect of which partial extension of macromolecules may take place, and
(2) a change in the nature of a solvent, which results in a change of equilibrium conformation of macromolecules.

These factors were mentioned soon after the first experimental works on liquid crystalline state of polymers were published. For example, Frenkel'[78] considered the effect of these factors by studying the change in the statistical flexibility parameter f introduced by Flory in his analysis of flexible chain polymers[53]. Frenkel' proposed the following equation to describe the effect of the type of the solvent on the flexibility of a polymer chain:

$$f = f_0 \exp\left(-\frac{\Delta\chi v_1}{kT}\right) \cdot \left[1 - f_0 + f_0 \exp\left(-\frac{\Delta\chi v_1}{kT}\right)\right]^{-1}$$

where the subscript "o" indicates a more flexible conformation, and v_1 is the volume fraction of a solvent.

In a similar way, the effect of an external field extending a macromolecule may be expressed in terms of the effective flexibility parameter in a stationary external field:

$$f = f_0 \exp\left(-\frac{\Delta\varepsilon}{kT}\right) \cdot \left[1 - f_0 + f_0 \exp\left(-\frac{\Delta\varepsilon}{kT}\right)\right]^{-1}$$

where $\Delta\varepsilon$ signifies the additional external energy.

The change in f to the value below the critical ($f_0 = 0.63$) may result in the liquid crystalline state appearing in a polymeric system which in the absence of external field or without a change in the nature of a solvent is in the isotropic state. This principle of equivalence of inherent and induced rigidity was additionally discussed in the review article by Frenkel' [79].

Miller et al. [55] have shown that the variation of f for semirigid-chain polymers does not change the general form of the diagram. However, with increasing of flexibility a narrow heterophase region is shifted toward higher concentrations and considerably broadened.

An attempt to estimate the effect of deforming external field on the phase diagram was made by Marucci et al. [80]. They introduced into the equation for the free energy and additional term proportional to the velocity gradient (Γ). As usual, the boundaries of phase transitions were determined by equating the chemical potentials of the coexisting phases. Figure 13 [81] shows the effect of the gradient Γ on the composition of the coexisting isotropic, v_2^*, and anisotropic, v_2^{**}, phases.

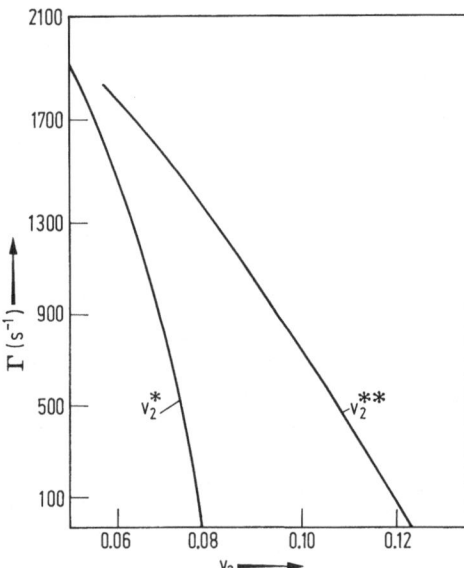

Fig. 13. The effect of the longitudinal velocity gradient Γ on the composition v_2^* and v_2^{**} of the coexisting phases (according to [81])

Valenti et al. [82] noted a transition to the mesophase under the effect of the applied external field for those semirigid-chain polymers which do not exhibit the liquid crystalline transition under normal conditions. A moderately concentrated solution of polyterephthalamide of p-aminobenzylhydrazide (X-500) in dimethyl sulphoxide does not exhibit the transition into the liquid crystalline state, which can be explained by a relatively low axial ratio for the macromolecules. The persistence length for this polymer is estimated being equal to 50 Å. At the same time, according to the data from the literature, high-modulus fibres have been obtained from the solutions of this polymer, which can be connected only with the appearance of the liquid crystalline state in the process of fibre formation. The authors of [82] believe that this transition begins already during the flow of the solution and called this phenomenon the flow-induced transition. Apparently, the rigidity of this polymer is close to the critical limit above which a transition to the liquid crystalline state becomes possible. It is interesting to note that X-500 forms a nematic phase in sulphuric acid solutions in the static state, i.e. without flow [83]. Consequently, the liquid crystalline state for the solutions of rigid-chain polymers may be induced both by a mechanical stress (flow) and by a choice of an appropriate solvent in which the rigidity of macromolecules increases (see also the work by Hartzler and Morgan [84]).

Naturally, the flow induced transition into the liquid crystalline state cannot be stable after the external field has been removed. Therefore, we must look for the explanation of the effect of fibre strengthening in the transition of the system to a highly concentrated state during spinning, when the above-mentioned persistence length already ensures the formation of the mesophase.

3.3 Kinetics of Phase Transitions in the Systems Involving Rigid-Chain Polymers

In most of the published papers concerning the transitions of polymers to the liquid crystalline phase, the equilibrium state is mainly considered, while intermediate (nonequilibrium) states are studied insufficiently. However, in general polymers exhibit slow phase transitions, especially those which are connected with the appearance of the ordered phase (polymer crystallization). A similar phenomenon takes place also for the transitions to the liquid crystalline state. Flory [53] paid attention to this effect in his early work while considering a spontaneous elongation of acetate cellulose films being treated with media which cause swelling (a decrease of the glass-transition point). When the films are formed by a rapid evaporation of a volatile solvent from a thin layer of the solution, cellulose acetate macromolecules have no time to acquire the equilibrium conformation and to form an ordered system, which is caused by a relatively high rigidity of the polymer. In order to acquire the minimal volume of the system, macromolecules must go over to a conformation with chain bendings. If, however, such a nonequilibrium film is subjected to swelling (e.g., in the aqueous phenol solution) or to heating, i.e. if it is transferred to the region above the glass-transition point, the equilibrium conformation of macromolecules is recovered and according to the principle of the minimum of the free energy, the liquid crystalline ordering must appear, since the rigidity of cellulose acetate chains lies within the limits for which such an ordering is possible (persistence length is equal to $75 = 100$ Å). The result of this ordering is a spontaneous elongation of the film in the direction given by a slight preorientation [85, 86].

Spontaneous elongation should be treated as an indirect indication of the transition of a system from a nonequilibrium disordered (amorphous) state to the equilibrium liquid crystalline state. Similar effects are also observed for other rigid-chain polymers which, according to the conditions of their processing, initially are in a nonequilibrium amorphous state. For example, Kalashnik et al. [87] have shown that during heating the fibres of poly(p-phenylene-1,3,4-oxadiazole) to the temperatures exceeding the glass-transition point (~ 300 °C) their spontaneous elongation takes place (Fig. 14). For the fibres based on flexible chain polymers obtained in similar condition, heating usually leads not to elongation but to contraction as a result of relaxation of internal stresses appearing during spinning of the fibre.

According to the nucleation mechanism, the rate of appearance of a new phase must decrease in the following order: amorphous phase > liquid crystalline phase > crystalline phase. This can be explained by the fact that for nuclei of the amorphous phase to appear, any fluctuation of particles, whose dimension exceed a certain critical value, is sufficient. At the same time, the nuclei in the liquid crystalline phase appear only at a more strict ordering of molecules, and for the appearance of nuclei of the crystalline phase a still more strict mutual arrangement of molecules is required.

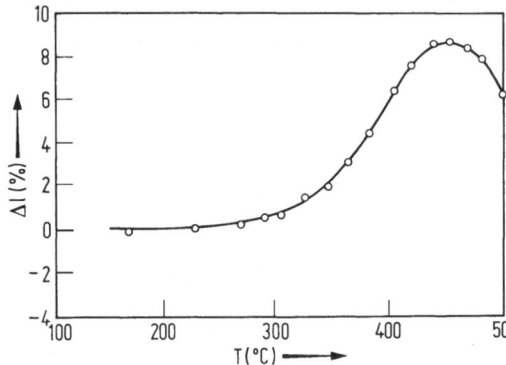

Fig. 14. Temperature dependence of deformation for poly(phenylene-1,3,4-oxa-diazole) fibres (according to [87])

According to the principle of mutual independence of individual types of phase equilibrium [59], it should be expected that upon a change of thermodynamic conditions in the initial monophase solution, first the equilibrium with the separation into amorphous phases is established, this equilibrium being unstable with respect to other types of phase equilibrium, and only after that the transition to the stable equilibrium takes place. As an example, we can consider the case of a gradual transition from the unstable liquid crystalline equilibrium to the stable equilibrium with the formation of a crystallosolvate for a PBA-sulphuric acid system, which we discussed earlier [88,89]. In the lower left part on Fig. 15 there are the particles of liquid crystalline phase which transforms, via the isotropic phase (dark background) into a crystallosolvate (spherulites in the upper part of the figure). The process is completed by a total disappearance of the liquid crystalline phase and by the establishment of the equilibrium between the isotropic solution and the crystallosolvate, which corresponds to the region I + CS on Fig. 12.

Figure 16 shows a schematic diagram of phase transformations for rigid-chain polymers separated from isotropic solutions by introducing a nonsolvent into the system (this is a usual method of obtaining fibres and films) (cf. [58]). The initial isotropic solution with the polymer concentration v_2^0 and the value of the Huggins-Flory parameter χ^0 is in the monophase region. The critical concentration of the transition into liquid crystalline state for this system is v_2^*. When a nonsolvent is introduced, i.e. when χ is increased up to the value >0.5 (χ'), two routes of the phase transition are possible: the separation into two amorphous phases with the concentration v_2^a for the phase enriched in a polymer (metastable state) along curve 1, or the separation along curve 2 with the formation of the liquid crystalline phase v_2^{lc} (stable state). If the nonsolvent is introduced sufficiently rapidly, initially the equilibrium between the two amorphous phases may be established (along curve 1). This equilibrium proves to be relatively stable from the point of view of kinetics of the high concentration and a correspondingly high viscosity (phase v_2^a). When the solvent is removed (e.g., by drying), the system may retain the amorphous nonequilibrium state

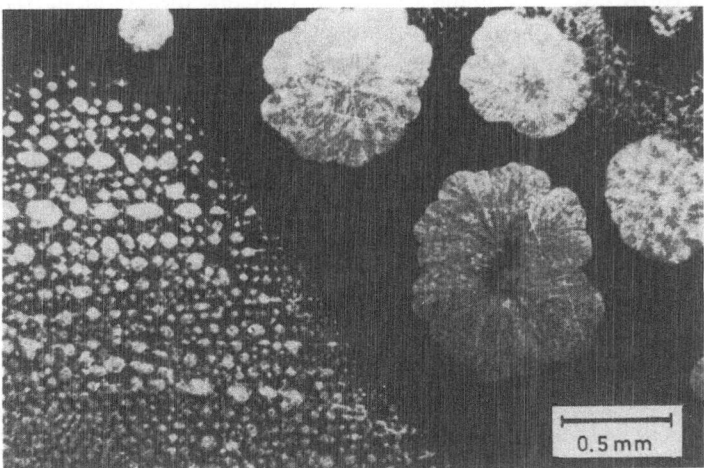

Fig. 15. Intermediate stage of the transition from the metastable equilibrium between the liquid crystal and the isotropic solution to the stable equilibrium between the polymeric crystallosolvate and the isotropic solution in a PBA-sulphuric acid system. The photograph is obtained in crossed polaroids (according to [88, 89])

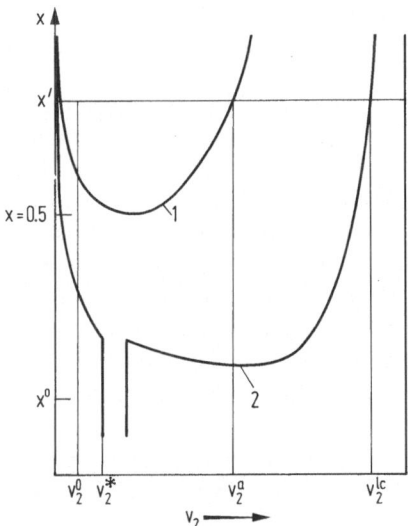

Fig. 16. The combination of the two types of phase equilibrium: liquid (curve 1) and liquid-crystalline (curve 2) (according to [58])

($v_2 = 1$). The transition (partial or complete) into the liquid crystalline state occurs only after the system is heated above the glass-transition point. For real polymeric systems with semiflexible chains, the liquid crystalline state in the initial solution often is not realized, so the formation of nonequilibrium amorphous polymer upon the introduction of a nonsolvent is quite probable.

It is not excluded that this mechanism is observed during the formation of fibres from X-500 [82]. The authors of this work pointed out that when the fibre was heated to 250–300 °C, its spontaneous elongation took place. Note that to attain higher orientation of a polymer in a fibre, it is necessary not only to transfer it to the liquid crystalline state but also to orient the liquid crystalline domains formed along the axis of the fibre. This orientation of the domain in which the macromolecules have been already mutually ordered requires not too high a draw ratio (the theoretical value must be <2). Indeed, experiments [82] have shown that at the draw ratio of 1.53 to 1,70 the modulus (E) and the tensile strength (σ) of the fibre at thermal treatment increase, which can be seen from the table compiled according to the results of this work.

The Variation of the Modulus and the Tensile Strength of X-500 Fibre at Thermal Treatment

T, °C	E, GPa	σ, GPa
initial fibre	14.6	0.56
200	28.0	0.88
250	48.0	0.96
300	53.0	1.10
350	59.0	1.30
400	67.0	1.30

The sharp increase in the modulus (by more then 4.5 times) at such a small draw ratio is not typical for flexible chain polymers. It is connected with the transition of the polymer into the ordered liquid crystalline state.

The problem of transition of semirigid-chain polymers into the liquid crystalline state is associated with a number of limitations that should be overcome. The obtaining of initial anisotropic solution for these polymers is complicated by the fact that because of a low degree of asymmetry of macromolecules, their transition to the liquid crystalline state is possible only at very high concentrations. But in this case the viscosity of the system becomes so high that the very establishment of such a transition and homogenization of the system (i.e. uniform distribution of a polymer and a solvent) become very difficult. In some cases a transition of a polymer from a nonequilibrium amorphous state to the equilibrium liquid crystalline state by heating is limited by a high glass-transition temperature which is often close to the temperature of intensive thermal degradation of the polymer.

The general scheme of the transition of semirigid-chain polymers from the non-equilibrium amorphous state to the liquid crystalline or crystalline state is given in [90].

Polymer crystallization may also complicate the transition from the amorphous state to the intermediate liquid crystalline state. Even small crystalline fractions in a polymer prevent free movement of macromolecules [91]. This is the case, for example, with cellulose. Cellulose fibres, obtained via cellulose xanthate (viscose), in the process of forming partially crystallize. Although the amorphous fraction in these fibres is large (up to 70–75%), at a short-term heating of the fibres above the glass-transition temperature (240–260 °C) only slight self-elongation of fibres is observed, which can be attributed to the transition to the liquid crystalline state at the expense of the amorphous fraction [92].

Recently, cellulose and its derivatives attracted researchers' attention again. Besides the work by Aharoni [93], who confirmed the appearance of the liquid crystalline state for cellulose acetate mentioned above, a number of other works appeared in which liquid crystalline state was established for hydroxypropyl cellulose [94–96]. Finally, Chanzy et al. [97,98] have obtained the results indicating the possible transition to the mesophase of cellulose itself, and not only its derivatives.

At the present time the range of rigid-chain polymers has considerably broadened. Now it includes helix-forming polypeptides, para-aromatic polyamides and polyesters, cellulose and its derivatives, polyisocyanates, and some other polymers.

4 Concluding Remarks

The rapid growth of the number of publications concerning polymeric liquid crystals indicates that we should expect the appearance of new fundamental studies on the transition of rigid- and semirigid-chain polymers into this state. The range of moderately concentrated solutions for these polymers is studied sufficiently well, while the development of the methods of establishing the liquid crystalline state in superconcentrated systems and in pure polymers with semirigid chains, as well as the analysis of kinetics of phase transitions, are the subject for further theoretical and experimental works.

The behavior of polymeric systems under compression (high pressures) has also been studied insufficiently. Till now the equilibrium under normal pressure has been mainly considered. The experiments on the formation of the ordered state of polyethylene under the pressure above 3–4 kbar revealed that there is much to be done in this field to extend the concept to polymers in general (see [99,100]).

Thus, the range of problems associated with phase transitions in polymers considerably broadens. If earlier the appearance of the liquid crystalline state for polymers was considered as an interesting exception, at the present time this phase state acquires in physics and chemistry of polymers equal rights with amorphous and crystalline states.

5 References

1. Elliot, A. E., Ambrose, E. J.: Discuss. Faraday Soc. 9, 246 (1950)
2. Robinson, C.: Trans Faraday Soc. 52, 571 (1956)
3. Robinson, C.: Tetrahedron 13, 219 (1961)
4. Robinson, C.: Mol. Cryst. 1, 467 (1966)

5. Kalmykova, V. D., Kudrjavzev, G. I., Papkov, S. P., Volokhina, A. V., Iovleva, M. M., Mil'kova, L. P., Kulichikhin, V. G., Bandurjan, S. I.: Vysokomol. Soedin. *B13*, 707 (1971)
6. Papkov, S. P.: Chim. Volokna *15*, No. 1, 3 (1973)
7. Papkov, S. P., Iovleva, M. M., Mil'kova, L. P., Antipova, R. V., Ghouchberg, S. S., Kudrjavzev, G. I., Volokhina, A. V., Kalmykova, V. D.: Vysokomol. Soedin. *B15*, 357 (1973)
8. Papkov, S. P., Bandurjan, S. I., Iovleva, M. M.: Vysokomol. Soedin. *B15*, 370 (1973)
9. Papkov, S. P., Iovleva, M. M., Mil'kova, L. P., Kalmykova, V. D., Volokhina, A. V., Kudrjavzev, G. I.: Vysokomol. Soedin. *B15*, 757 (1973)
10. Sokolova, T. S., Jefimova, S. G., Volokhina, A. V., Kudrjavzev, G. I., Papkov, S. P.: Vysokomol. Soedin. *A15*, 2505 (1973)
11. Papkov, S. P., Kulichikhin, V. G., Kalmykova, V. D.: J. Polym. Sci., Polym. Phys. Ed., *12*, 1753 (1974)
12. Kwolek, S. L.: USA Patents 3, 600, 350 (1971), 3, 671, 542 (1972)
13. Morgan, P. W.: ACS Polymer Preprints *17* (1), 47 (1976)
14. Kwolek, S. L., Morgan, P. W., Schaefgen, J. R., Gulrich, L. W.: ACS Polymer Preprints *17* (1), 53 (1976)
15. Panar, M., Beste, L. F.: ACS Polymer Preprints *17* (1), 65 (1976)
16. Schaefgen, J. R., Foldi, V. S., Loguello, F. M., Good, V. H., Gulrich, L. W., Killian, F. L.: ACS Polymer Preprints *17* (1), 69 (1976)
17. Bair, T. I., Morgan, P. W., Killian, F. L.: ACS Polymer Preprints *17* (1), 59 (1976)
18. Papkov, S. P., Kulichikhin, V. G.: Liquid Crystalline State of Polymers (russ.). Khimija, Moscow, 1977
19. Liquid Crystalline Order in Polymers. A. Blumstein, Ed. Academic Press, New York, 1978
20. Rigid Chain Polymers. Synthesis and Properties. G. C. Berry and C. E. Sroog, Eds.: J. Polym. Sci., Polym. Symp., *65* (1978)
21. Samulsiki, E. T., DuPré, D. B.: Polymeric Liquid Crystals, in: Advances in Liquid Crystals, *4*, G. H. Brawn, Ed., Academic Press, New York, 1979
22. Papkov, S. P.: Vysokomol. Soedin. *A19*, 3 (1977)
23. Platé, N. A., Shibajev, V. P.: Comb-like Polymers and Liquid Crystals (russ.). Khimija, Moscow, 1980
24. Morgan, P. W.: Macromolecules *10*, 1381 (1977)
25. Kuhn, W.: Koll. Z. *68*, 2 (1934)
26. Tsvetkov, V. N.: Vysokomol. Soedin. *A19*, 2171 (1977)
27. Birshtein, T. M.: Vysokomol. Soedin. *A19*, 34 (1977)
28. Erman, B., Flory, P. J., Hummel, J. P.: Macromolecules, *13*, 484 (1980)
29. Hummel, J. P., Erman, B., Flory, P. J.: Macromolecules, *13*, 476 (1980)
30. Prozorova, G. E., Pavlov, A. V., Iovleva, M. M., Antipova, R. V., Kalmykova, V. D., Papkov, S. P.: Vysokomol. Soedin. *B18*, 111 (1976)
31. Tsvetkov, V. N.: Vysokomol. Soedin. *A20*, 2066 (1978)
32. Tsvetkov, V. N.: Vysokomol. Soedin. *A21*, 2606 (1979)
33. Arpin, M., Strazielle, G.: Polymer *18*, 597 (1977)
34. Papkov, S. P.: Vysokomol. Soedin. *B24*, 869 (1982)
35. Iovleva, M. M., Papkov, S. P.: Vysokomol. Soedin. *A24*, 233 (1982)
36. Gallot, B. in: Liquid Crystalline Order in Polymers, A. Blumstein, Ed., Academic Press, New York, 1978
37. Onsager, L.: Ann. N.Y. Acad. Sci. *51*, 697 (1949)
38. Kayser, R. F., Raveche, H. J.: Phys. Rev. *A17*, 2067 (1978)
39. Flory, P. J.: Proc. Roy. Soc. (London), *A234*, 73 (1956)
40. Flory, P. J., Ronca, G.: Mol. Cryst. Liq. Cryst. *54*, 284, 311 (1979)
41. Flory, P. J.: J. Polym. Sci. *49*, 105 (1961)
42. Miller, W. G., Wu, C. C., Santec, G. L., Rai, J. H., Gaebel, K. G.: Pure Appl. Chem. *38*, No. 1–2, 37 (1974)
43. Iovleva, M. M., Papkov, S. P., Mil'kova, L. P., Kalmykova, V. D., Volokhina, A. V., Kudrjavzev, G. I.: Vysokomol. Soedin. *B18*, 830 (1976)
44. Flory, P. J.: ACS Polymer Preprints *17* (1), 46 (1976)
45. Flory, P. J., Abe, A.: Macromolecules, *11*, 1119 (1978)
46. Abe, A., Flory, P. J.: ibid., 1122 (1978)

47. Flory, P. J., Frost, R. S.: ibid., 1126 (1978)
48. Frost, R. S., Flory, P. J.: ibid., 1134 (1978)
49. Flory, P. J.: ibid., 1138 (1978)
50. Flory, P. J.: ibid., 1141 (1978)
51. Jackson, W. J., Kuhfuss, H. F.: J. Polym. Sci., Polym. Chem. Ed. *14*, 2043 (1976)
52. MacFarlane, F. E., Nicely, V. A., Davis, T. G. in: Contemporary Topics in Polymer Science, vol. 2, E. M. Pearce and J. R. Schaefgen, Eds., Plenum Press, New York, 1977
53. Flory, P. J.: Proc. Roy Soc. (London) *A234*, 60 (1956)
54. Krigbaum, W. R., Salaris, F.: J. Polym. Sci., Polym. Phys. Ed. *16*, 883 (1978)
55. Miller, W. G., Rai, J. H., Lee, E. D., in: Liquid Crystals and Ordered Fluids, *2*, R. Porter and J. Johnson, Eds. Plenum Press, New York, 1974. p. 233
56. Werbowyi, R. S., Gray, D. G.: ACS Polymer Preprints *20* (1), 102 (1979)
57. Nakajama, A., Hayashi, T., Ohmori, M.: Biopolymers 6, 973 (1968)
58. Papkov, S. P. in: Contemporary Topics in Polymer Science, vol. 2. E. M. Pearce and J. R. Schaefgen, Eds., Plenum Press, New York, 1977, p. 97
59. Papkov, S. P.: Phase Equilibria in Systems Polymer-Solvent (russ.). Khimija, Moscow, 1981
60. de Gennes, P. G.: The Physics of Liquid Crystals, Clarendon, Oxford, 1974
61. Khokhlov, A. R.: Vysokomol. Soedin. *A21*, 1981 (1979)
62. Grosberg, A. Ju., Khokhlov, A. R.: Advances in Polymer Science *41*, 53 (1981)
63. Flory, P. J.: 179th ACS Nat. Meet., Houston, 1980, Abstr. Pap.
64. van der Waals, J. D.: Lehrbuch der Thermostatik, Bd. II, J. A. Barth, Leipzig, 1927
65. Baird, D. G.: in Advances in Liquid Crystals, *4*, G. H. Brawn, Ed., Academic Press, New York, 1978, p. 237
66. Wissbrun, W. F.: J. Rheology *25*, 619 (1981)
67. Papkov, S. P., Kulichikhin, V. G., Malkin, A. Ja., Kalmykova, V. D., Volokhina, A. V., Gudim, L. J.: Vysokomol. Soedin. *B14*, 244 (1972)
68. Kulichikhin, V. G., Platonov, V. A., Braverman, L. P., Belousova, T. A., Poljakov, V. F., Shablygin, M. V., Volokhina, A. V., Malkin, A. Ja., Papkov, S. P.: Vysokomol. Soedin.: *A18*, 2656 (1976)
69. Onogi, S., Asada, T.: in Rheology, vol. I, G. Astarita, G. Marucci and L. Nicolais, Eds., Plenum Press, New York, 1980, p. 127
70. Kolzov, A. I., Belnikovich, N. G., Gribanov, A. V., Papkov, S. P., Frenkel', S. Ja.: Vysokomol. Soedin. *B15*, 645 (1973)
71. Platonov, V. A., Kulichikhin, V. G., Papkov, S. P., Litovchenko, G. D., Belousova, T. A., Shablygin, M. V.: Vysokomol. Soedin. *A18*, 221 (1976)
72. Aikawa, Y., Minami, N., Sukigaru, M.: Mol. Cryst. Liq. Cryst. *70*, 115 (1980)
73. Patel, D. L., DuPré, B.: J. Polym. Sci., Polym. Lett. Ed., *17*, 299 (1979)
74. Andrejeva, I. N., Khanin, Z. S., Romanko, O. I., Volokhina, A. V., Iovleva, M. M., Kalashnik, A. T., Papkov, S. P., Kudrjavzev, G. I.: Vysokomol. Soedin. *B23*, 89 (1981)
75. Iovleva, M. M., Smirnova, V. N., Khanin, Z. S., Volokhima, A. V., Papkov, S. P.: Vysokomol. Soedin. *A23*, 1867 (1981)
76. Iovleva, M. M., Papkov, S. P.: Vysokomol. Soedin. *A24*, 233 (1982)
77. Papkov, S. P.: Vysokomol. Soedin. *B24*, 109 (1982)
78. Frenkel', S. Ja.: J. Polym. Sci., Polym. Symp. *44*, 49 (1974)
79. Frenkel', S. Ja.: Pure Appl. Chem. *38*, 117 (1974)
80. Marucci, G., Sarti, G. C., in: Ultra-High Modulus Polymers, A. Ciferri and J. M. Ward, Eds., Applied Science Publishers, London, 1979
81. Ciferri, A.: Intern. J. Polymeric Mater. *6*, 137 (1978)
82. Valenti, B., Alfonso, G. C., Ciferri, A., Giordani, P., Marucci, G.: J. Appl. Polym. Sci. *26*, 3655 (1981)
83. Morgan, P. W.: J. Polym. Sci., Polym. Symp. *65*, 1 (1978)
84. Hartzler, J. D., Morgan, P. W., in: Contemporary Topics in Polymer Science, vol. 2. E. M. Pearce and J. R. Schaefgen, Eds., Plenum Press, New York, 1977, p. 19
85. Majury, T. G., Willard, H. J.: Simposio intern. di chimica macromoleculare, Roma, 1955, p. 354
86. Belnikevich, N. G., Bolotnikova, L. S., Ediljan, E. S., Brestkin, Ju. V., Frenkel', S. Ja.: Vysokomol. Soedin. *B20*, 37 (1978)

87. Kalashnik, A. T., Volokhina, A. V., Semjenova, A. S., Kuznetsova, L. K., Papkov, S. P.:
    Chim. Volokna *19*, No. 4, 51 (1977)
88. Papkov, S. P., Iovleva, M. M., Bandurjan, S. I., Ivanova, N. A., Adrejeva, I. N., Kalmykova,
    V. D., Volokhina, A. V.: Vysokomol. Soedin. *A20*, 658 (1978)
89. Papkov, S. P.: Vysokomol. Soedin. *B21*, 787 (1979)
90. Kalashnik, A. T., Papkov, S. P.: Vysokomol. Soedin. *A23*, 2302 (1981)
91. Kalashnik, A. T., Papkov, S. P., Kudrjavzev, G. I., Bobrovnitskaja, N. I., Mil'kova, L. P.:
    Vysokomol. Soedin. *A22*, 2000 (1980)
92. Kalashnik, A. T., Papkov, S. P.: Vysokomol. Soedin. *B18*, 455 (1976)
93. Aharoni, S. M.: Mol. Cryst. Liq. Cryst. (lett.) *56*, 237 (1980)
94. Werbowyi, R., Gray, D.: Mol. Cryst. Liq. Cryst. (lett.) *34*, 37 (1976)
95. Werbowyi, R., Gray, D.: Macromolecules, *13*, 69 (1980)
96. Bheda, J., Fellers, J. F., White, J. L.: J. Appl. Polym. Sci. *26*, 3955 (1981)
97. Chanzy, H., Dube, M., Marchessault, R. H.: J. Polym. Sci., Polym. Lett. Ed. *17*, 219
    (1979)
98. Chanzy, H., Peguy, A., Chaunis, S., Monzie, P.: J. Polym. Sci., Polym. Phys. Ed. *18*, 1137 (1980)
99. Yasuniwa, H., Enoshita, R., Takemura, T.: Japan. J. Appl. Phys. *15*, 1421 (1976)
100. Eljashevich, G. K., Frenkel', S. Ja.: in Orientational Phenomena in Solutions and Melts of
     Polymers (russ.). A. Ja. Malkin and S. P. Papkov, Eds., Khimija, Moscow, 1980

N. A. Platé (guest editor)
Received March 18, 1983

Liquid Crystal Polymers with Flexible Spacers in the Main Chain

Christopher K. Ober
Xerox Research Centre, Mississauga, Ontario, Canada L5K 2L1

Jung-Il Jin
Department of Chemistry, Korea University, Seoul 132, Korea; Qifeng Zhou,
Chemistry Department Beijing University, Beijing, China (P. R. C.)

Robert W. Lenz
Chemical Engineering Dept., University of Massachusetts, Amherst, Ma 01 003, USA

This review is be concerned primarily with the synthesis and properties of such polymers which show thermotropic behaviour.

The literature on this subject can be organized in several ways, but the most straightforward method is to compare the effect of polymer structure on physical properties. These properties are principally a function of three structural variables: the type of mesogenic unit and the length and type of flexible spacer. This review is, therefore, organized to treat these variables, starting with a discussion of the effects of the mesogenic unit on polymer properties and followed by a consideration of the changes in polymer behaviour caused by the use of different flexible spacers. The methods used for the characterization of LC polymers are also discussed where appropriate.

1 Introduction 104

2 Mesogenic Groups in Main Chain LC Polymers 105
   2.1 Mesogenic Groups Containing Ester Links 109
   2.2 Mesogenic Groups Containing Azo and Azoxy Links 112
   2.3 Mesogenic Groups Containing *Trans*-Vinylene Links 113
   2.4 Mesogenic Groups Containing Imino Links 113
   2.5 Mesogenic Groups Based on Rings Other Than p-Phenylene 113
   2.6 Comparison of Mesogenic Group Effects 114
   2.7 Effect of Substituents 117

3 Flexible Spacers in Main Chain LC Polymers 120
   3.1 Polymethylene Spacers 121
   3.2 Poly(ethylene oxide) Spacers 125
   3.3 Polysiloxane Spacers 127
   3.4 Effect of Substituents in the Spacer 128

4 Main Chain LC Copolymers 130

5 Characterization of LC Polymers 132
   5.1 Microscopy 132
   5.2 X-Ray Diffraction 135
   5.3 Thermal Analysis 137
   5.4 Rheological Properties 140

6 Field Effects on LC Polymers 143

7 References 143

Advances in Polymer Science 59
© Springer-Verlag Berlin Heidelberg 1984

1 Introduction

Liquid crystalline compounds are remarkable because of their ability to show spontaneous anisotropy and readily induced orientation in the liquid crystalline state. When polymers are processed in the liquid crystalline state, this anisotropy may be maintained in the solid state and can readily lead to the formation of materials of great strength in the direction of orientation. A particularly important example of the use of this property for polymers is in the formation of fibers from aromatic polyamides which are spun from shear oriented liquid crystalline solutions [1]. Solutions of poly(benzyl glutamate) also show characteristics of liquid crystalline mesophases, and both of these types of polymers are examples of the lyotropic solution behaviour of rigid rod polymers which was predicted by Flory [2].

In contrast to these lyotropic mesogenic materials, which require the presence of a solvent in order to produce a mesophase, thermotropic mesogens are those compounds which exhibit a mesophase in the melt state at temperatures above the crystalline solid state and before the formation of an isotropic melt [3]. Mesophase-forming polymers may possess either one of the two basic structures shown in Fig. 1; the polymer may either contain the mesogenic group (the part of the polymer molecule which is responsible for liquid crystallinity) directly in the main chain, or the mesogen may be present as a pendant group in the side chain.

Side chain liquid crystalline, LC, polymers and cross-linked LC polymers, with mesogenic units in the cross-link, have been the subject of a recent review [4] and constitute the major topic of two books [5,6]. Those polymers that contain a flexible spacer linking the main chain to the pendant mesogenic group often show a greater ability to form a mesophase than polymers lacking a spacer, when other factors are the same [7]. The spacer serves to dissociate the disorder of the main chain from the greater order of the mesogenic groups and to decouple the motions of the mesogenic moiety from those of the polymer backbone.

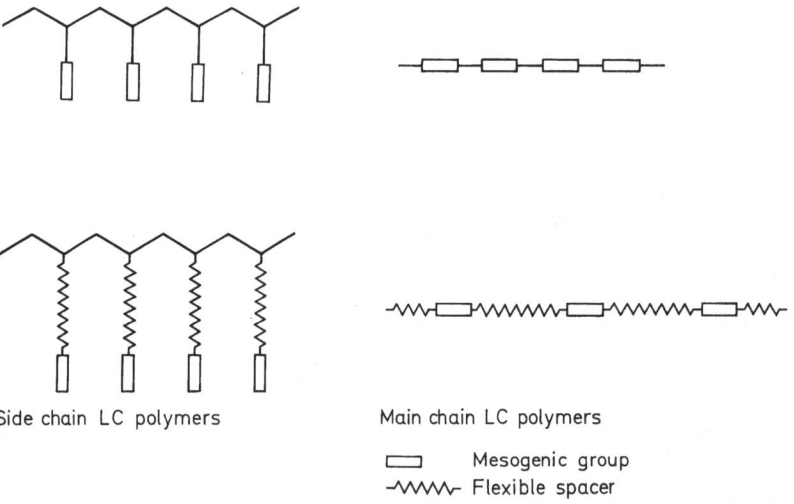

Side chain LC polymers   Main chain LC polymers

           ☐   Mesogenic group
   ⌇⌇⌇⌇ Flexible spacer

Fig. 1. Thermotropic Polymers with Either Main Chain or Side Chain Mesogenic Units

Flexible spacers have also been extensively used to separate mesogenic groups placed in the main chain from each other, and the backbone flexibility achieved by this approach has the effect of markedly improving solubility and lowering transition temperatures in contrast to those of the rigid rod polymers of comparable composition [8]. However, thermotropic rigid rod polymers may be modified in other ways to improve their processability, and such polymers are receiving increased interest for potential fiber applications because they can be processed by melt spinning techniques as opposed to the necessity of using highly active solvents as required in the solution spinning of polyamide fibers [9].

In addition to the use of a flexible spacer, other methods can be used to reduce the transition temperatures of main chain LC polymers to a reasonable range as compared to the extremely high melting points which are characteristic of rigid rod polymers; these include: (1) the use of substituents, (2) the addition of an element of dissymmetry to the main chain by copolymerizing mesogenic units of different shapes, referred to as "frustrated chain packing", and (3) copolymerizing non-linear, non-mesogenic units [10]. The term "frustrated chain packing" can refer to any mechanism that preserves the linearity and stiffness of the rigid rod polymer chain but hinders crystal or liquid crystal packing. The use of non-linear groups in the main chain destabilizes the liquid crystal state by reducing the effective persistence length of the chain and decreases the melting point in the usual manner of copolymers.

All of these methods can produce a reduction of the transition temperatures without adversely affecting the range of mesophase stability, but the greater part of the literature concerning LC polymers with main chain mesogenic units has been concerned with the use of flexible spacers for this purpose.

2 Mesogenic Groups in Main Chain LC Polymers

The term "mesogenic group", for the purposes of this review, refers to the part of the polymer chain that is composed of the rigid, linear segments and the atoms or functional groups which link them together in a linear array. It is this part of the polymer chain that ultimately determines whether or not the polymer will be liquid crystalline, within what range the transition temperatures will occur for a thermotropic polymer, and what type of mesophase can be formed.

In Fig. 2 is a schematic representation of the general structure of a liquid crystalline polymer having both a linear rigid mesogenic group (such as the p-phenylene ring) and a flexible spacer in the main chain. The mesogenic group must consist of at least two aromatic (or cycloaliphatic) rings connected in the *para* positions by a short rigid link which maintains the linear alignments of the aromatic rings. In this manner a rigid element is formed which has an overall length that is substantially greater than the diameter of the aromatic group. The linking groups used in LC polymer systems have included imino, azo, azoxy, ester, and *trans*-vinylene groups, and a direct link between aromatic rings may also be used, such as in biphenyl and terphenyl units. The mesogenic group may include combinations of two or more aromatic rings or cycloaliphatic rings to which the links are attached to maintain linearity; e.g., 1,4-phenylene, 2,6-naphthalene, and *trans*-1,4-cyclohexylene rings. Table 1 shows representative structures of these combinations of rings and links, as well as flexible spacers, which have been used as the basis for the preparation of thermotropic LC polymers.

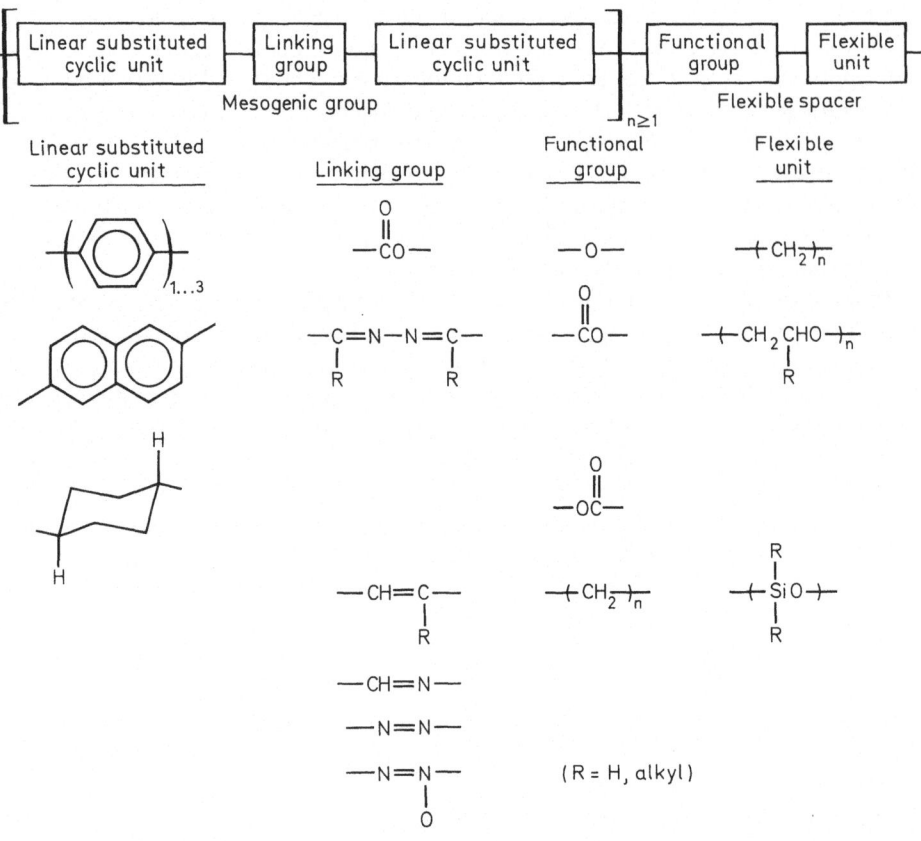

Fig. 2. General Structure of Mesogenic Units in Main Chain Polymers

Table 1. Liquid Crystalline Polymers with Main Chain Mesogenic Groups and Flexible Spacers

	Mesogenic unit	Flexible spacer	n	Ref.
1)	⬡-C=N-N=C-⬡ (CH₃, CH₃)	—OCO+CH₂+ₙ-OCO—	6,8,10,12	23,24)
2)	"	—OCO+CH₂CH₂O+ₙ-CO—	2,3,4	58)
3)	"	—OC+CH₂+ₙ-CO—	8,10,12	25,27)
4)	⬡-C=CH-⬡ (CH₃)	"	6,8,10,12	25,63,64)
5)	"	—OCO+CH₂CH₂O+ₙ-CO—	2,3,4	58)
6)	⬡-CH=CH-⬡	—C-O+CH₂+ₙ-O-C—	5,6,10	22)
7)	"	—OC+CH₂+ₙ-CO—	10	20)
8)	⬡-CH=N-⬡	"	10	20,73)

Table 1. (continued)

	Mesogenic unit	Flexible spacer	n	Ref.
9)			4,6,8,10,12,16	19)
10)	//	—O—(CH₂CH₂O)ₙ—	2,3,4	37)
11)			4,6,8,10,12,16	19)
12)	//		2,3,4	37)
13)	//		10	19)
14)		//	10	47)
15)			5,6,8,12 10	30) 20,73)
16)	//		2,4,6	22)
17)		//	2,4,5,6,10	22)
18)	//		2,3,4,10	22)
19)		—O—(CH₂)ₙ—O—	2...11	12,30)
20)	//		4,6,7,8	38)
21)	//	—O—(CH₂CH₂O)ₙ—	4	38)
22)	//		2,3,5	39)
23)		—O—(CH₂)ₙ—O—	10	17)
24)		//	10; X = H, Cl Y = H, Cl, Me, Br	14)
25)	//		X = H Y = Br	60)

Table 1. (continued)

	Mesogenic unit	Flexible spacer	n	Ref.
26)		$-O+CH_2\overline{)}_n O-$	10	17)
27)		//	9,11 10	30) 17)
28)	//	$-O+CH_2\overline{)}_3 +SiO\overline{)}_n Si+CH_2\overline{)}_3 O-$ (Me, Me / Me, Me)	2,3	40)
29)		//	2,3,5	39)
30)		//	2,3	40)
31)		$-O+CH_2\overline{)}_n O-$	5...10	14)
32)	//	$-O-CH_2-Si-O-Si-CH_2-O-$ (Me, Me / Me, Me)	--	60)
33)		$-OC+CH_2\overline{)}_n CO-$	3...12,14, 20	30,31)
34)	//	$-O+CH_2\overline{)}_n O-$	4,5,6,8,9	30)
35)	//	//	copolymers of 2...10,12 and 6,8,10	13,28,29)
36)		$-CO(CH_2)_n OC-$	2...10,12	33)
37)	//	$-CO+CH_2 CH_2 O\overline{)}_n C-$	2,3,4,8,13	34)
38)		$-CO+CH_2 CH_2 O\overline{)}_n C-$	3	55)
39)	//	$-C-O+CH_2\overline{)}_n O-C-$	2,6,10	33,55)

	Mesogenic unit	Flexible spacer	n	Ref.
40)	⬡—N=CH—⬡—CH=N—⬡	||	2,6,12	56)
41)	⬡—CH=N—⬡—⬡—N=CH—⬡	||	12	57)

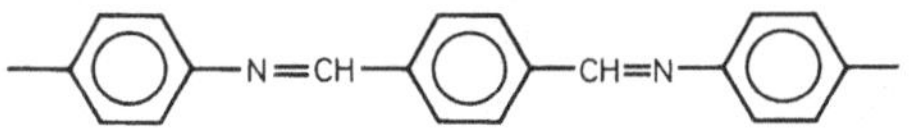

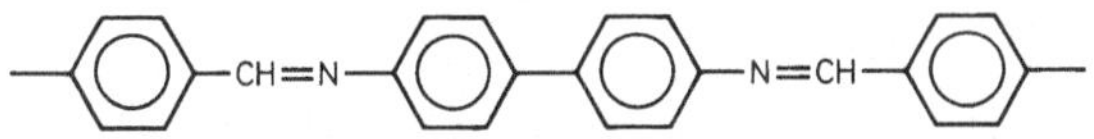

2.1 Mesogenic Groups Containing Ester Links

LC polymers of the ester type have been the most widely synthesized because they can be prepared by low temperature solution polymerization methods and melt transesterification reactions from available or readily attainable monomers. High temperature ester interchange polymerization, however, is somewhat limited in usefulness because it often leads to polymer structures of poorly defined sequence [11].

The ester mesogenic structures reported to date have generally contained either two or three aromatic units as illustrated by Polymers 15 to 39 in Table 1. These structures have been synthesized using mainly terephthalate, p-oxybenzoate and hydroquinone units in either their unsubstituted or substituted forms. Linkage to the flexible spacer in the terminal *para* positions of these mesogenic group ester dyads and triads may be by either ester or ether linkages, and of course the former link can be formed in either of two orientations, as is the case in Polymers 6 and 7 of Table 1. Indeed, the arrangement of this ester linkage has a noticeable effect on the mesophase properties of these polymers even when the same spacer length and type are used.

As shown below, there are essentially four types of symmetrical, ester triad mesogenic structures, and three types of ester dyad mesogenic structures:

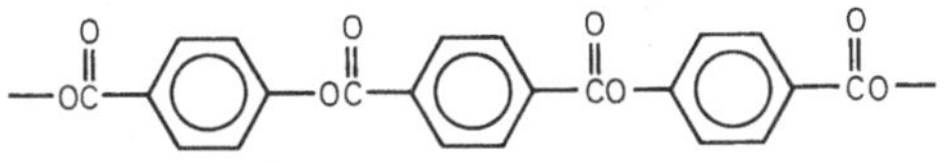

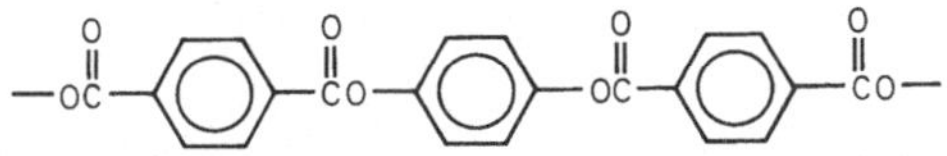

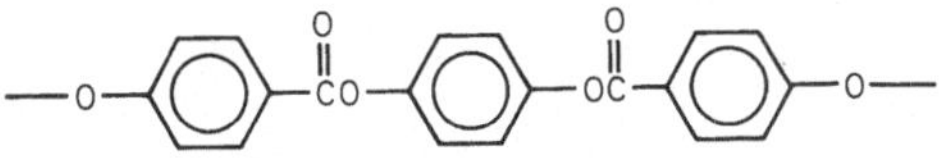

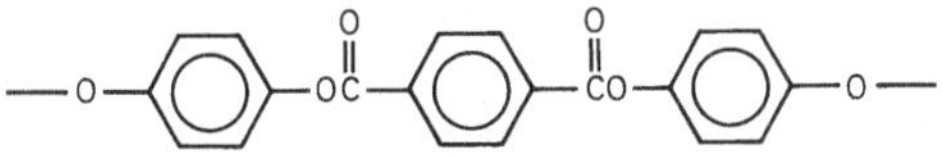

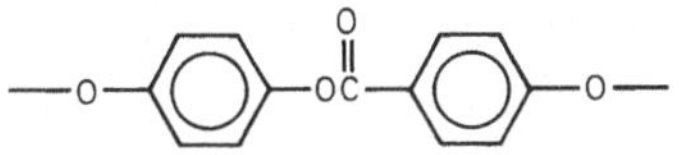

These mesogenic group structures will be referred to by the abbreviations in brackets in which O is a p-oxybenzoate group, H is a hydroquinone group and T is a terephthaloyl group. A fairly large number of polymers with these mesogenic structures have been prepared and are reported in the literature [12-15]. They have been characterized by thermal and optical methods, and in several cases by wide angle X-ray diffraction, WAXD, directly on the LC melt. Three of the mesogenic structures (OTO, OHO and OH) are discussed in some detail in the later section of this review devoted to polymethylene spacers.

Table 2 lists the transition temperatures and temperature ranges of mesophase stability, ΔT, of the dyad and triad ester polymers in which a common flexible spacer is present, the decamethylene spacer. This spacer gives polymers with melting temperatures in a range very convenient for study, and for this reason a large number of polymers have been prepared using this spacer with different mesogenic groups.

The melting temperature is, of course, dependent on the sample history and molecular weight, so polymer samples within such a series must be heat treated in the same manner and be of sufficiently high molecular weights for comparisons to be possible. For the latter, studies on both the OTO triad [101] and on an azoxy polymer [43] show that polymer transition temperatures level off at fairly low molecular weights, as

Table 2. Triad and Dyad Polymers with a Decamethylene Spacer

	Link	Approx. Length	T_m (°C)	T_i (°C)	ΔT	$[\eta]$ dl/g	Meso-phase[a]	Ref.
a) Triad Polymers								
1) OTO	ester	19Å	220	267	47	0.45	S	33)
2) THT	ester	19Å	230	265	35	0.60	N	44)
3) HTH	ether	18Å	236	265	29	0.22	N	14)
4) OHO	ether	18Å	237	294	57	0.18	N	14)
b) Dyad Polymers								
1) OH	ester[b]	13	175	260	85	0.25	N	31)
2) OH	ether[c]	11.5	185	212	27	0.60[f]	N	13)
3) TO	ester[c]	13	140	—[d]	—	0.20	N	55)
4) TO	ester[b]	13	140	—[e]	33	0.20	N	55)

[a] designations: S — forms smectic phase, N — forms nematic phase;
[b] dyads are randomly arranged in a head-to-tail, tail-to-head sequence;
[c] dyads are arranged in an exactly alternating head-to-tail, tail-to-head sequence;
[d] no mesophase observed;
[e] monotropic liquid crystalline phase formed on cooling at 133 °C;
[f] estimated

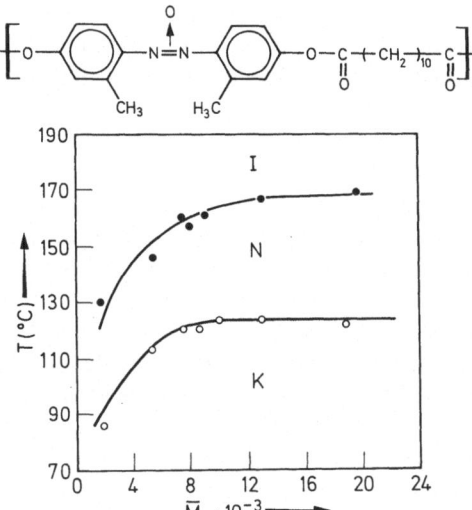

Fig. 3. Triad and Dyad Mesogenic Units [43]

shown for the latter in Figure 3. Two of the triad mesogenic groups in this series, the OTO and OHO groups, reportedly give polymers which can form a smectic mesophase [12, 15], but the other two triads form polymers that have only nematic mesophases [14]. The OTO triad polymer appears to form a smectic A phase, but the OHO triad polymer forms only a poorly defined smectic phase. There is some controversy in the latter case about the true mesophase present because only one of three research groups working with that structure observed a smectic mesophase [14, 16].

An interesting feature of the triad polymers is evident from the data in Table 2, which contains the melting temperatures, T_m, and the temperature at which the liquid crystal melt phase is changed to an isotropic melt phase, T_i. The latter is often referred to as the clearing temperature because the LC melt is generally opaque. All of the triad polymers in this table melt and clear at approximately the same temperature, as might be expected from their similar structures. The polymer with the OTO triad mesogenic group melts at 220 °C while the others melt at about 235 °C. The polymers in which the link from the spacer to the mesogenic unit is an ether function have slightly higher melting temperatures than those with ester links.

The different behaviours of the four polymers are not easily explained. Structurally, the triads have similar bond angles so the geometries of the different mesogenic group should be quite similar. The dipole moments would vary according to the arrangement of the ester groups, but the manner in which this arrangement affects the structure and stability of the mesophase cannot be easily predicted.

The use of more rigid or more flexible central units in the triad has also been studied by several research groups for Polymers 23 to 29 of Table 1. When a p,p'-biphenylene ring replaces the p-phenylene ring, both melting and clearing temperatures increase greatly as would be expected, and when a bulkier aromatic ring is used, such as binaphthylene or a substituted p-phenylene, the melting temperatures drop, especially in the latter case [17]. The range of mesophase stability of the substituted polymers, however, is also greatly reduced, while the increased rigidity of the biphenylene ring formed a more stable mesophase. Presumably, the bulkiness of the substituted aromatic rings led to hindered mesophase packing. The higher entropy of clearing, ΔS_i, and reduced ΔT of the polymer based on phenylhydroquinone can be attributed to an interlocking effect between polymer chains.

The inclusion of smaller substituents on the central aromatic ring of such triad ester mesogens has also been investigated, including methyl, methoxy, cyano, and several halide groups [18]. As expected, the presence of substituents reduces the melting temperature because of the effect of molecular dissymmetry on crystal lattice packing, but the clearing temperature is not necessarily decreased, and as a result, ΔT is often increased in such monosubstituted polymers. These results will be discussed in more detail in the section on the effect of substitution on mesophase behavior.

Polymers with dyad ester mesogenic units consist of two basic structures depending on whether the terminal functional groups of the dyad are hydroxy or acid groups; i.e., the OH and TO dyads, respectively in Table 2. All of these polymers, if liquid crystalline, possess nematic mesophases, but some show monotropic behavior. Both the ranges of mesophase stability and the melting temperatures are reduced in comparison to those of the comparable triad polymers. These polymers contain only one intermesogenic unit linkage and, therefore, the dyad shows a directional sense in contrast to the symmetrical triad mesogenic units.

The OH dyad has been incorporated into polymers where the flexible spacer was either a diacid or a diol as illustrated by Polymers 33–35 of Table 1, which contain an ester and an ether group between the spacer and mesogenic group. Polymers containing the OH dyad had higher melting temperatures and wider values of ΔT than the polymers with TO dyads as can be seen, for example, by comparing the properties of Polymers b1 and b4 of Table 2. Polymer 1 has a large ΔT while Polymer 4 is monotropic and has a lower melting temperature, although both polymers are nematic. This result is opposite to the general observation that terminal carboxy groups on the mesogenic unit favor a more stable mesophase.

The behavior of Polymers b3 and b4 of Table 2 is rather interesting. Both are of the same structure, the only difference being that in Polymer 3 the central ester linkage is placed in the chain in an alternating fashion. This small regularity in structure causes a loss of the mesophase while the compositionally identical polymer with a random placement of the ester link has a monotropic nematic phase.

2.2 Mesogenic Groups Containing Azo and Azoxy Links

Most of the mesogenic groups in LC polymers based on azo or azoxy links contain two aromatic rings, and a fairly large number of polymers with azoxy-containing mesogenic groups and two p-phenylene rings have been prepared and form stable liquid crystal phases as illustrated by Polymers 11 to 14 in Table 1. However, when carbonyl groups terminate this mesogenic structure, the range of mesophase stability is limited and either monotropic or non-LC polymers are formed [19,20]. By simply reversing the direction of the ester group (which would change the extent of conjugation), the melting temperature is raised only slightly (molecular weight differences may be important in these examples), but ΔT is increased substantially. These latter polymers showed nematic phases and no monotropism.

Structurally similar polymers with azo links, Polymers 9 and 10, formed no meso-phases when the rigid aromatic unit contained terminal carbonyl groups, but these formed nematic mesophases when the ester link was reversed [19]. The transition temperatures of the azo and azoxy polymers with the same spacer were quite close to each other. Substitution of methyl groups on the phenylene rings of these polymers reduced melting temperatures without adversely affecting nematic mesophase stability [21].

2.3 Mesogenic Groups Containing *Trans*-Vinylene Links

The *trans*-vinylene or *trans*-stilbene type of mesogenic structure forms polymers with properties similar to those of the azoxy polymers discussed above [20, 22]. This similarity of behaviour is anticipated from consideration of the nearly identical geometry and size of the mesogenic structures. The connection between the mesogenic group and the spacer is again important, and reversal of the orientation of the ester group linking the spacer to the mesogenic structure has a large effect on ΔT. Polymers with mesogenic groups having terminal ester units as in Polymer 7 have greater ther-mal stabilities than for those with reversed terminal ester units as in Polymer 6, and one author has observed smectic behavior for the former [20]. This observation, which is similar to the results for the azo and azoxy polymers, was based on polymers with decamethylene spacers.

Polymers with methyl substitution on the central linkage in the stilbene mesogenic group have been prepared, Polymers 4 and 5, and show strictly nematic mesophases [60]. The methyl group seems to act mostly to inhibit the existence of smectic order, but otherwise does not influence the mesophase formation or thermal stability very much.

2.4 Mesogenic Groups Containing Imino Links

Another type of unsaturated mesogenic structure, which is based on a structure in between an azo- and vinylene-type, is that in which a Schiff base unit links two phe-nylene groups. This unit is similar in geometry to the three previous mesogenic units, the azo, azoxy and stilbene groups [20]. The only reported example of this structure, however, was Polymer 8 with a decamethylene spacer, and it was quite similar in properties to the polymers just discussed. It also possessed a smectic phase, and the mesogenic unit had terminal carboxy groups.

Extension of the linking groups by use of a double Schiff base (Polymers 1–3 of Table 1) creates a linking group with an extended conjugated structure. A series of polymers based on this mesogenic unit was prepared by Roviello and Sirigu [23]. The diphenolic mesogenic form was used and reacted with either diacids, or diformyl compounds [24]. Only relatively long spacer lengths of more than 8 methylene units were studied, and the resulting polymers formed only nematic mesophases [25].

2.5 Mesogenic Groups Based on Rings Other Than p-Phenylene

The terphenyl unit is a very rigid, extended mesogenic structure because of the direct linkage between phenylene rings, Polymers 17 and 18 of Table 1. A series of polymers with this group was prepared from the p,p'-carboxy terminated terphenyl, and these generally formed smectic mesophases if either an alkylene or a polyoxyethylene spacer was used [22]. Both the smectic A and C phases were observed in these polymers, and a smectic E phase was also suspected.

The shorter biphenyl group has been used with alkyl spacers, Polymers 15 and 16 of Table 1, to give both smectic and nematic polymers depending on the direction of the linking ester group [20,22,26]. Again, it has been found that if the mesogen is a phenol ester the nematic state is favored while a mesogenic ester based on aromatic acids gives exclusively smectic polymers. In this case and in that of the terphenyl polymers, the range of mesophase stability of the smectic state is substantial.

Several polymers have made use of mesogenic structures with more than one type of linking groups. Examples of these polymers are those containing either biphenylene or bicyclohexylene groups flanked by oxybenzoate groups [39], Polymers 27 to 29 of Table 1. A third type of mixed mesogenic group, Polymer 30, is one composed of a Schiff base linkage between two phenylene groups, which are in turn surrounded by oxybenzoate groups [40]. Only the first example has been used with a polymethylene spacer, but the latter two were prepared with polysiloxane spacers. As would be expected, the polymers with biphenylene central groups had higher melting and clearing temperatures, but the clearing point was the one more affected so that a broader mesophase range was obtained than for that of the comparable polymer with a hydroquinone central group, Polymer 19.

2.6 Comparison of Mesogenic Group Effects

Several different structural factors influence the properties of the mesophase in these polymers, including dipolar effects, the planarity and rigidity of the mesogenic unit, and its length-to-width ratio among others. These factors are difficult to quantify, either absolutely or relatively, but some idea of their influences can be obtained by comparing the properties of polymers with different mesogenic units when combined with the same flexible spacer. This comparison has already been made for the dyad and triad esters in Table 2, and in this section it will be extended to other types of liquid crystalline polymers which contain a common decamethylene spacer.

The data for this comparison are given in Table 3. Polymers in which ether groups link the spacer to the mesogenic unit are listed in the first part of the table, while those with ester linking groups are located in the latter part. Both groups of polymers have ten methylene groups in the spacer, but the presence of the carbonyl in the ester adds two more carbon atoms to the repeat structure of the polymer chain compared to the ether-linked polymers. Nevertheless, in both series there is a simple and similar correlation between the length of the mesogenic group and the melting temperature. As observed by Strzelecki, the longer the aromatic segment the higher the melting temperature. Other factors such as the rigidity and polarity of the mesogenic group are also important, but these have a much greater effect on the clearing temperature of the mesophase than on the melting point.

When different classes of polymers are compared, however, the correlation with length is not so straightforward. The polymers with OTO and THT ester triads of Table 2 which have mesogenic lengths of approximately 19Å, melt and clear at lower

Table 3. Comparison of Properties of Liquid Crystalline Polymers Containing Different Mesogenic Groups and a Decamethylene Spacer

Mesogenic Structure	Approx. Length, Å	Tm (°C)	Ti (°C)	ΔT °C	Ref.
1)	18	151(N)	168	19	17)
2)	22	224(N)	248	24	17)
3)	22	258(N)	354	96	30, 17)
4)	15.5	256(S)	311	55	22)
5)	13.5	210	—	0	19)
6)	13.5	200	—	0	19)
7)	11.0	154(?)	160	6	22)
8)	13.5	225(N)	245	20	19)
9)	13.5	197(N)	200	3	22)
10)	13.5	216(N)	265	49	47)
11)	13.5	118(N)	162,5	42.5	47)

Table 3. (continued)

Mesogenic Structure	Approx. Length, Å	Tm (°C)	Ti (°C)	ΔT °C	Ref.
12)	13.5	196(N)	218	22	47)
13)	16.0	217(N)	242	25	23, 24)
14)	13.5	200(N)	222	22	25)

temperatures than the shorter but more rigid terphenyl polymer, Polymer 4 of Table 3. The shorter azo and azoxy mesogenic groups give polymers with lower melting temperatures and no mesophase, Polymers 5 and 6, while the shorter biphenyl polymer has an even lower melting temperature and a mesophase, Polymer 7. Apparently the planarity and rigidity of the terphenyl and biphenyl groups are very important for mesophase formation.

Comparing, finally, the polymers made from diphenolic compounds and α,ω-alkane diacids, the Schiff base structure in Polymer 13 of Table 3 and the methyl-stilbene structure in Polymer 14 gave polymers with about the same range of mesophase stability, but the polymer with the shorter stilbene unit has a lower melting temperature. However, the azo and azoxy polymers, Polymers 8 and 10, which have mesogenic structures of the same size as the stilbene unit possess higher melting temperatures, and the azoxy polymer has a wider range of mesophase stability. Evidently, the polarity of the azoxy polymer, and probably that of the azo polymer too, is very important in both aspects of mesophase behavior.

Substitution on the azoxy polymers with methyl groups affects the length-to-diameter ratio of the mesogenic unit, and leads to a reduction of transition temperature; the degree of which depends on the position of the methyl group. Similar results are observed with the phenyl substituted ester triad of Polymer 1 in Table 3, which gives a polymer with both a lower melting temperature and ΔT than those of the unsubstituted triad polymer. Likewise, the polymer with a binaphthol central unit, Polymer 2, melts at a lower temperature than the polymer with a central biphenyl unit, Polymer 3, and it also possesses a narrower ΔT. Steric repulsion by the substituents combined with interference in the coplanarity between phenyl rings could be the structural factors behind such observations.

The effect of reversing ester groups, which has already been briefly discussed, should be mentioned again here. In the case of the azo and azoxy polymers made from carboxy-terminated flexible spacers, polymers with stable mesophases were formed. In contrast, polymers made from oxy terminated spacers had no mesophase. While this behavior is an extreme case, when comparisons are possible it seems that polymers made with the carboxy-terminated spacers almost always have wider ranges of mesophase stability. No explanation can be offered at the present time for these narrower thermal stability ranges of the mesophases obtained from polyesters based on oxyterminated flexible spacers.

There is as yet no clear-cut relationship between the molecular structure of the mesogenic unit and the type of mesophase it forms, but several generalizations can be made. Gray and Winsor [3] have divided these factors into how the molecular structure: (1) is conducive to liquid crystal formation, (2) affects the thermal stability of the mesophase, and (3) favors the occurrence of smectic *versus* nematic or cholesteric liquid crystals.

Because the molecules are geometrically anisotropic, a stage-wise breakdown of the crystal might occur. For example, as a smectogen melts, the interlayer forces may also act to preserve the structure. As well as having the appropriate cohesive forces, the molecule must be sufficiently rigid that in the molten state flexing does not occur. For this reason, most substances that give mesophases are of aromatic character. By substitution in the *para* position and connecting at least two rings by a rigid group, a rod-like mesogenic group is formed.

Many of these cases fall into the general category shown in Fig. 2, with a flexible spacer present and n can be a value of one or more. When the ester group is the connecting link between the phenylene rings, the mesogenic group maintains its overall linearity, but the ester group is more mobile than other linking groups, so a less thermally stable mesophase is formed. When n is two, this linking group can form polymers with a monotropic mesophase.

If more than two groups are connected (n > 2), then the liquid crystalline properties are enhanced, especially if conjugation or direct linkage of benzene rings is involved. This type of structure leads to both higher transition temperatures and increased LC stability as demonstrated by a wider temperature range for mesophase behavior; that is, a greater thermal stability of the mesophase. Linkages through the *ortho* or *meta* positions will destroy liquid crystallinity by destroying molecular linearity, and linear structures with substituents in these positions usually have lower LC thermal stability. In general, mesophase behavior can be affected greatly by modifying both rigidity and the length of the mesogen, which is generally described in terms of the length-to-width ratio of the molecular structure of the mesogen.

Liquid crystalline behavior can also be modified by the addition of lateral substituents, as mentioned above. The substituents can have either one of two effects: either (1) they increase intermolecular attractions through changes in molecular dipole moment and/or polarizability, or (2) they act to separate the long axes of the mesogens through steric interactions. Of course, these factors depend on the size and type of substituent and the mesogen involved. The position of the substituent is also important, and in certain cases an added substituent can either create a mesogen from a non-mesogen or destroy the liquid crystallinity [3]. The following section of this review discusses the effects of both changing the size of the substituent and its polarity on the properties of the resulting liquid crystalline polymers.

2.7 Effect of Substituents

Extensive property modifications occur, as described above, when substituents are introduced into the mesogenic units of a main-chain liquid crystalline polymer.

From a review of the results of studies on low molecular weight liquid crystals [3], it is known that substituents can act to reduce the coplanarity of adjacent mesogenic groups and increase the diameter or decrease the axial ratio of the mesogens. In addition, the ordering of neighboring mesogenic groups becomes more difficult because the side groups force the molecules apart in order to meet their own steric requirements. However, an increase in dipolar interactions between neighboring polar substituents can also occur and it may be helpful, while the other effects may be detrimental, to the thermal stability of the liquid crystalline phase.

One of the earliest studies on the effect of substituents on lowering the melting and clearing temperatures of main-chain, thermotropic LC polymers is that described in patents issued to the du Pont Company [41]. Their polymers were prepared in the melt by transesterification of dicarboxylic acids and diacetates of chloro-, bromo- or methyl-substituted hydroquinones. The use of these ring substituted hydroqui- nones in random copolymers resulted in depressing the melting temperatures of these polymers and allowed them to be melt spun in their liquid crystalline state. Later, Millaud and coworkers [42] found that the high melting temperatures of aromatic LC polyazomethines could be lowered by introducing a methyl group onto the phenylene- diamine rings.

The effect of substituents on the thermal properties of a series of aromatic azoxy polymers is shown in Table 3, Polymers 10–12, in which it is seen that the introduction of methyl groups into the mesogenic units resulted in a depression in both the melting point and the clearing temperature of the polymer. As can be seen from these data, the substitution position in the mesogenic unit is also important, but it will be necessary to have much more experimental results before the positional effect may be evaluated systematically.

Because of the difficulty of introducing various substituents into the mesogenic units of LC polymers, few such systematic studies have been made, but several series of substituted polymers were synthesized and examined in our laboratories.

The first series of substituted homopolymers had the following general structure [14]:

where X was either H, CH$_3$, Cl, Br or phenyl group. With the introduction of any of these substituents, the melting point of the unsubstituted polymer was depressed by at least 60–70 degrees, but the isotropization temperature was less affected unless a very large side group, such as the phenyl group, was used. Therefore, the temperature ranges of the stable mesophases in this polymer series were broadened by the presence of the smaller substituents.

Another polymer series studied was the alkyl substituted poly(1,4-phenylene terephthalate) [43]. The melting temperature of the parent polymer in this series is so high (> 500 °C) that a stable mesophase is not able to exist without thermal degra- dation occurring. The alkyl substituents were found to be effective in lowering the transition temperatures of the polymer, and it was found that both the melting and clearing temperatures decreased with increasing side chain length (from the linear hexyl group to the linear dodecyl group).

The alkyl groups were found to affect the clearing temperature in a different manner than the melting temperature in the following polymer series [44]:

$$\left[OC-\bigcirc\!\!\!\bigcirc-CO-\bigcirc\!\!\!\bigcirc-OC-\bigcirc\!\!\!\bigcirc-CO+CH_2\!\!\:\,)_{10} \right]$$

As shown in Fig. 4, by replacing a hydrogen atom with a methyl group, $R = CH_3$, a remarkable decrease in melting point, from 231 to 154 °C, occurred. This temperature further decreased in an approximately odd-even, zig-zag fashion when the alkyl group was changed from methyl to hexyl. With further increase in the length of side groups, the melting point increased slightly and then leveled off. It is likely, therefore, that a further increase in the length of the pendent alkyl group beyond eight carbon atoms does not produce any additional steric effect that can interfere with the molecular packing in the solid state, and indeed the constant melting point for the polymers with longer alkyl groups may be that resulting from the crystallization of the side chain alkyl groups themselves rather than the polymer main chains.

On the other hand, the clearing temperatures of the polymers in this series decreased steadily with increasing length of the substituent, although the contribution of each additional methylene unit to the depression of this transition temperature became smaller, as shown by the gradually decreased slope of the curve in Fig. 4. This phenomenon may be explained as the result of the gradually decreased contribution of each additional methylene unit to the molecular diameter (as defined by Gray [3]) of the mesogenic units.

If the alkyl group is lengthened to more than 6 carbon atoms, the clearing temperature of this particular polymer series either may have been depressed so much that

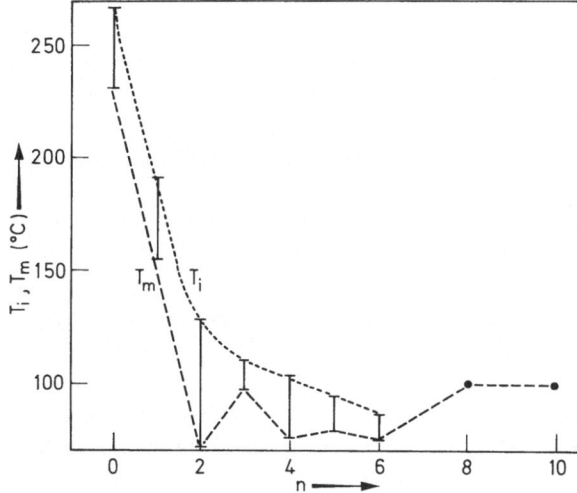

Fig. 4. Dependence of Transition Temperature on Molecular Weight for Polymer 14 of Table 1 [44]

it was even lower than the melting point or the polymer was incapable of forming a liquid crystalline mesophase; such was the case also for polymers with octyl- and decyl-substituents.

Because of the zig-zag manner of the change in the melting point and the steady decrease in the clearing temperature with increasing length of the alkyl groups, the temperature range of the stable mesophase of these alkyl-substituted polymers changed also in a nearly zig-zag fashion. As seen in Fig. 4, the ethyl-substituted polymer had the lowest melting point (71 °C) and the widest mesophase range (56 °C). Hence, substituents can depress both the melting point and the isotropization temperature, but they do not necessarily narrow the temperature range of the stable mesophase, especially when the alkyl group is small.

Substituents other than alkyl groups were also studied [44]. It was found that highly polar substituents (e.g., —CN, —NO$_2$) were very effective in depressing both the melting and clearing temperatures of these polymers. This depression may be again considered to be partly the result of the steric effects which limited the molecular packing efficiency in both the crystal and the liquid crystal state, but the polar effect is believed to be important also, and the latter resulted in a higher clearing temperature for the bromo, cyano and nitro-substituted polymers. These substituents are all larger in size than the methyl group, but the clearing temperatures of polymers with these substituents were found to be higher than those of the methyl-substituted polymer. The methoxy-substituted homopolymer in this series was found to be non-liquid crystalline (possibly it is monotropic) with a melting point of 158 °C.

Copolymerization of two monomers with different substituents was used as another approach for depressing the transition temperatures of the polymers [44]. For example, a copolyester prepared from an equimolar mixture of cyano- and methoxyhydroquinone and 1,10-decane-*bis*(p-chloroformylbenzoyloxy)decane had a melting point of 133 °C and a clearing temperature of 177 °C, while the corresponding temperatures of the cyano-substituted homopolymer were 157 and 219 °C.

3 Flexible Spacers in Main Chain LC Polymers

In the previous section, the mesogenic unit was defined as the part of the polymer chain that is composed of the aromatic or cycloaliphatic segment. Simply for convenience, we have defined the remainder of the polymer repeating unit as the flexible spacer. What is important, however, is the different structural nature of the spacer and the mesogenic unit, and unfortunately the separate roles of these two groups are not yet fully understood.

Most of the subject polymers that have been studied to date have contained flexible spacers composed of polymethylene chains. The monomers or precursors for these spacers are readily available as either diols or diacids or as dihalides, and this availability has led to the preparation of several homologous series of thermotropic polymers with different mesogenic units. Indeed, the first reported main chain, thermotropic LC polymer contained a polymethylene flexible spacer [27].

More recent studies on the effect of flexible spacers now include the use of either poly(ethylene oxide) or polysiloxane segments. Each such spacer brings with it certain unique characteristics to the final polymer, and it is clear that the role of the spacer is more complex than simply that of decoupling the mesogenic groups. These spacers will be discussed in the following sections along with their substituted forms.

3.1 Polymethylene Spacers

The effect of increased polymethylene flexible spacer length on the properties of main chain LC polymers has been very well documented in a number of publications [25-30]. Essentially three effects can be observed in all such series: (1) reductions of the transition temperatures (both melting points and clearing temperatures) with increased spacer length, (2) an even-odd relationship for the transition temperatures, in which polymers with an even number of atoms in the spacer generally have higher transition temperatures than those in the same series with an odd number of atoms, and (3) in some cases, a smectic mesophase is formed by polymers containing very long spacers (approximately eight atoms or more). All of these effects are illustrated in Fig. 5.

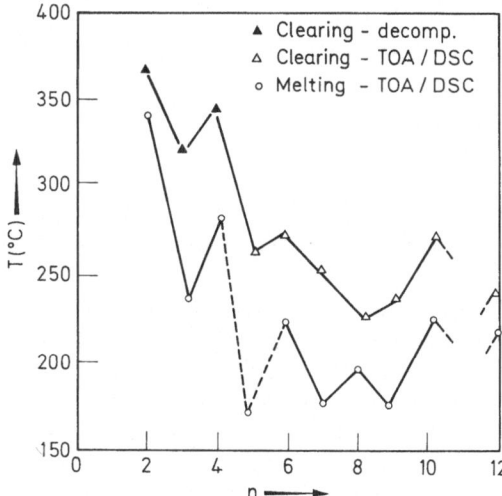

Fig. 5. Dependence of Transition Temperature on Spacer Length for a Series of Polymers Containing Mesogenic Group 36 of Table 1 [33]

The even-odd effect was first detected in the series of nematic polyesters prepared by Griffin and Havens, the mesogenic structure of which is listed as 35 in Table 1 [28]. They also found that the temperature range for mesophase stability decreased as the spacer length increased, and the copolymer with the longest spacers in their series (10 and 12 methylene groups) had the lowest transition temperatures [29].

Soon after the results for these polymers appeared in the literature, Strzelecki and Van Luyen reported the preparation of another homologous series based on the mesogenic structure 19 [30]. In the range of spacer lengths of 4–9 methylene groups, a decrease of transition temperature was again observed with increased spacer length, but moreover, both nematic and smectic mesophases were formed by polymers containing spacers which were 6 methylene units long and longer. The ability to form the more highly ordered smectic state as well as the nematic state with the longer spacers indicates that the greater flexibility of these longer chains allows the mesogenic units to align themselves into such highly ordered structures, or at least to retain that three-dimensional crystal-like order after melting.

Another series of polymers was prepared by reaction of the diacetoxy ester of mesogenic structure 33 in Table 1 with a series of α,ω-polymethylene diacids [31]. The
spacer length was increased from 3 to 12 methylene groups, and examples with spacers
of 14 and 20 repeat units were also prepared. These polymers had a mesogenic unit
that was shorter than that for the series just discussed (18Å vs 13Å), and this probably
led to the reported formation of only the nematic mesophase in this polymer series.

We have reported the synthesis of a homologous series of polyesters with mesogenic
structure 36 that was characterized by DSC, microscopy and wide-angle X-ray
diffraction [15,32]. In addition to the even-odd effect itself, it was also noted that the
polymers with odd-numbered spacers had greater ranges of mesophase stability
because the clearing transitions were not as sensitive to the even-odd effect as were
the melt transition. These polymers, like those of the triad series of Strzelecki and Van
Luyen, also possessed smectic mesophases for higher spacer lengths, but unlike their
series of polymers, only one mesophase could be observed for each polymer.

The entropies of clearing of these polymers, ΔS_i, also showed an even-odd relationship, but they demonstrated no particular trends [33]. Values for the entropy of clearing
were of the same order as those found for polymers that formed only nematic mesophases and which also contained polymethylene flexible spacers [21,25]. The thermodynamic properties of the mesophase will be discussed later. The values observed
seemed to be characteristic of those for polymers with polymethylene spacers in general, because other thermotropic polymers which contained poly(alkylene oxide)
spacers showed much lower values of ΔS_i even though they exhibited both nematic
or smectic mesophases [22,34]. Such results emphasize the very important role that
the flexible spacer plays in determining the nature of the mesophase.

Another series of closely related polymers based on structure 31 was also prepared
in our laboratory, and these showed analogous results to the previous series for polymers having 5 to 10 methylene groups in the spacer [14]. Again the polymers with
even-numbered spacers had lower ranges of mesophase stability, but the measured
values for the entropy of clearing showed amounts comparable to those of low molecular weight LC compounds having the same mesogenic structure. Usually, polymers
with polymethylene spacers show entropies of clearing that are much higher than
those for their low molecular weight counterparts [27].

Roviello and Sirigu have also recently prepared and characterized the even- and
odd-numbered members of several series of LC polymers [23]. They too discovered
an even-odd relationship in two of the three series of polymers investigated where
spacer lengths varied between 8 and 14 methylene groups [25]. The polymers with
mesogenic structures 1 and 3 in Table 1 showed the now typical effects of both higher
transition temperature for those with even -numbered spacers and a reduced range
of mesophase stability with increased spacer length. Interestingly, the polymer of
structure 2 which had a carbonate linkage between the spacer and mesogenic groups
showed a decrease in transition temperatures but an increase in mesophase stability
with increased spacer length. The widest temperature range of polymer mesophase
stability, in fact, occurred at a spacer length of 11 methylene units.

The entropies of clearing of these three series of polymers were determined by DSC
measurements and showed a remarkable even-odd relationship for the determined
values as shown in Fig. 6. Polymers of the series 1 mesogen showed a very weak
even-odd effect of values of approximately 3 cal mol^{-1} K^{-1}, while the other two

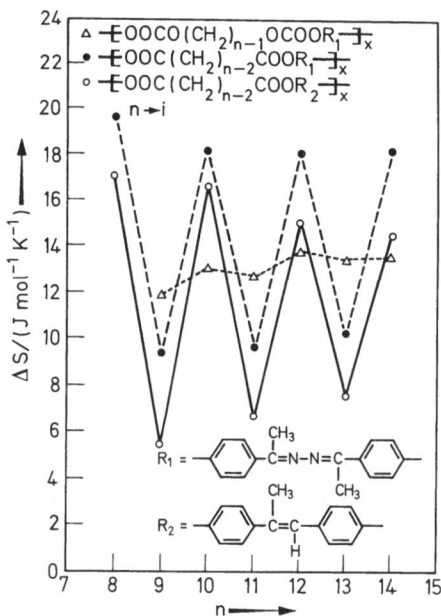

Fig. 6. Variation in the Entropy of Clearing with Spacer Length for Polymers 1, 2 and 4 in Table 1 [23]

mesogen series (3 and 4) fluctuated much more strongly between the values of 4 cal mol^{-1} K^{-1} and 2 cal mol^{-1} K^{-1}, with even-numbered polymers showing the higher values.

In similar work by Blumstein and Thomas, polymers of mesogenic structure 14 with high molecular weights were prepared, and these also showed the same relationships observed previously for transition temperatures and entropies of clearing [21]. In this case, a uniform increase was observed in the clearing entropies of both the odd and even series, and an additive value to ΔS_i of 0.36 cal mol^{-1} K^{-1} could be calculated for each methylene unit. From such values it was concluded that over 75% of the CH$_2$ groups in the spacer were in the *trans* conformation, thus providing evidence that the flexible spacer was in the extended state. However, no other evidence could be offered to confirm this supposition, and these results may pertain only to this series of polymers.

Several more examples of homologous series of LC polymers were studied in the investigations of Koide and coworkers [14]. The structures of these polymers are listed as Polymers 9 to 13 in Table 1. All showed the even-odd trend in melting temperatures, but only two series, the p,p'-azophenol and the p,p'-azoxyphenol polymers showed mesomorphic behavior over all spacer lengths covered. The p,p'-azobenzoates showed no mesophases. The p,p'-azoxybenzoates showed mesophase formation with spacers of only 5 and 7 methylene groups, and these polymers were monotropic. The observed mesophases were described as exhibiting the Schlieren texture on a polarizing microscope, but the type of mesophase was not identified.

The virtually universal even-odd behavior of both melting and clearing temperatures for liquid crystalline polymers having polymethylene spacers has not yet been explained theoretically, but similar trends have also been observed for low molecular weight LC compounds with alkyl terminal groups. Of course, odd-even effects are well known for the melting points of both low molecular weight compounds and polymers with polymethylene structures and are generally explained in terms of packing regularity in the crystal lattice.

Explanations of these types of trends have been based on changes in either molecular polarizability [65, 66] or conformational effects resulting from the lengthening of the alkyl end groups and passing from even to odd terminal chain lengths [67−69]. Most models assume that the mesogen behaves as a rigid cylinder, and any change in geometry caused by rotation of bonds that leads to violation of the cylindrical shape results in formation of the isotropic phase. That is, the explanation for a reduction of clearing temperature with increased spacer length is generally based on the effect of increasing the number of possible conformations to the long spacers with the resulting distortions of the cylindrical shape of the mesogen [66]. According to the rigid-rod theory, an all *trans* conformation, as illustrated in Fig. 7, favors the liquid crystalline state, so the increased addition of *gauche* bonds would reduce the mesophase order. At high enough temperatures, the large number of *gauche* conformations present in the terminal groups of low molecular weight LC compounds distort the cylindrical shape of the mesogen and destroy the liquid crystalline order [66]. The term "mesogen" in this discussion refers to the entire repeating unit or compound.

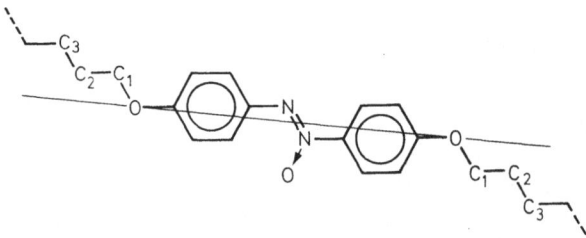

Fig. 7. Effect of Long Alkyl Terminal Groups on the Conformation of Low Molecular Weight Liquid Crystals [87]

The even-odd effect is induced by changes in the molecular polarizability of the mesogen in its normal and perpendicular components [65]. The polarizability along the molecular axis is greater than that perpendicular to the axis for even chains, but is about equal for odd chains if an all *trans* conformation is assumed. Stronger attractions exist between mesogens with even end groups, and consequently these compounds have higher clearing temperatures. A similar explanation is offered for analogous odd-even effects on polymer melting points.

Such factors lead to alternation of T_i, and an eventual damping of the effect itself. This interpretation is supported by an alternation of the order parameters that has been recently observed by ^{13}C-NMR [70] and that was predicted by Marcelja [71] and Pink [72] based on modified Maier-Saupe theory [59]. Their theoretical calculations have shown that the flexible end groups in LC compounds play an active role in the formation of the mesophase and that interactions between end groups and the mesogenic group cause an alternation of the order parameter, which in turn leads to the observed even-odd effects for ΔS_i.

Such reasoning could be logically extended to liquid crystalline polymers with the added consideration that rather than having free end groups, the mesogenic units act to constrain the spacers and therefore to reduce the number of possible conformations that the spacer groups might have. This rationale qualitatively explains the higher clearing temperatures observed for polymers when compared to model mesogens having end groups of lengths comparable to those of the flexible spacer in the polymer because higher energy (that is, higher temperatures) are required to destroy the polymer mesophases compared to model compound mesophases.

3.2 Poly(ethylene oxide) Spacers

Polymers of this type have been prepared with various mesogenic units connected by oligomers of poly(ethylene oxide), PEO, and it is possible in some cases to use spacers having a broad distribution of molecular weights instead of a monodisperse molecular species. It should be noted that each oxyethylene unit adds three bond lengths to the spacer, so the comparison of polymers with different types of spacers should be between spacers having the same numbers of atoms in the spacer chain.

Meurisse and coworkers studied a series of polymers based on the terphenyl mesogenic structure, 17 of Table 1, with spacers composed of the monomer, dimer, trimer, tetramer and decamer of PEO [22, 35]. From an initial melting temperature of 322 °C and clearing temperature of 393 °C for the polymer with monomer spacer, the series showed a sharp drop in both T_m and T_i for the dimer spacer, and the values finally leveled off with the decamer spacer at $T_m \sim 70$ °C and $T_i \sim 117$ °C. Polymers with polymethylene spacers of comparable lengths discussed above had approximately the same transition temperatures, and so the authors concluded that the oxygen in the PEO chains was stereochemically equivalent to the methylene group. The polymer with the PEO tetramer spacer was shown to be isomorphous with a smectic C reference compound [35]. The other polymers with PEO spacers showed typical smectic type textures when observed under the microscope, and the smectic morphology has recently been confirmed by X-ray diffraction [36]. The change in the entropies of clearing of this series was of the same order of magnitude as for polymers showing liquid crystalline behavior and having polymethylene spacers [22].

Iimura and coworkers synthesized azo and azoxy type polyesters, Polymers 10 and 12 of Table 1, containing PEO dimer, trimer and tetramer spacers and measured their thermal properties in addition to making observations by X-ray diffraction and on a polarizing microscope [37]. Many of the polymers showed focal conic textures that are associated with the smectic mesophase, but it was difficult to relate the types of mesophases for the polymers to their relative melting and clearing transitions. One finding was that T_i decreased with increasing spacer length, and T_m also decreased in the same manner as for non-liquid crystalline polymers of similar structure.

One of the most surprising observations was that the X-ray diffraction pattern of the azoxy polymers showed crystalline-like d-spacings in the liquid crystalline melt while in the isotropic region the polymers showed the expected amorphous halo. A possible explanation of such a highly ordered mesophase was the retention of much of the order of the solid in the molten state. A highly structured mesophase occurs in the smectic E state where directional and layered order are maintained in the melt in addition to some residual crystalline order. Low molecular weight compounds show highly structured X-ray diffraction patterns in the smectic E mesophase, and so a similar melt structure might be present in these polymers.

In order to study the effect that spacers have on mesophase behavior, Roviello and Sirigu prepared polycarbonates from the dimer, trimer and tetramer of PEO using benzalazine, Polymer 2, and methyl stilbene, Polymer 5, mesogenic units. The clearing and melting temperatures for each set of polymers followed the same general trend of decreasing with increasing spacer length. The clearing temperatures, T_i, were always higher for the polymers of structure 5 [54].

Both types of polymers showed enthalpies of clearing much lower than their polymethylene counterparts, and the values of ΔS_i observed were in the range of those for low molecular weight nematic liquid crystals. The low enthalpy of clearing was, therefore, an added support for the presence of a nematic state in the polymer melts.

The polymers with diethylene oxide spacers (PEO dimer) reportedly had, in addition to a nematic mesophase, a monotropic smectic mesophase that was detected by X-ray diffraction. No confirmation was available by microscopy because almost all polymers developed a homogeneous texture that gave no real detail.

The inspection of the X-ray diffraction pattern of the crystalline solid phase allowed the conclusion that the 21.8 Å spacing (the presence of which can be taken as proof of the smectic phase) was not due to residual solid material because of the absence of stronger reflections.

Skorokhodov and coworkers have reported the synthesis of a polymer having mesogenic structure 21 of Table 1 with a tetraethylene oxide spacer. The polymer was reported to melt at 185 °C and clear at 210 °C. The mesophase was said to be nematic, but except for this information and the method of preparation, very little additional information was reported for these polymers [38].

Recently, we have reported the synthesis and characterization of a series of polymers with a triad mesogenic structure, 37, using several different oligomers of PEO as spacers [34]. The molecular weights of the PEO spacer oligomers varied between 200 and 1000, and those spacers with molecular weights greater than 400 had rather large values of $\overline{M}_w/\overline{M}_n$. An interesting feature of this series of polymers was the presence of two mesophases for polymers having spacers based on the PEO dimer, trimer and tetramer.

Even though the melting and clearing temperatures were equivalent to those for the analogous polymers containing polymethylene spacers and the same mesogenic group, the mesophase properties were quite different. The entropies of clearing, for example, were much lower for the polymers with PEO spacers, and as in previous observations [24], were in the range of the values of ΔS_i for low molecular weight compounds. Microscopy studies of the polymers indicated that the higher temperature phase was nematic, and that the lower temperature phase was smectic, but because the latter phase was homeotropic, exact identification was difficult. X-ray diffraction patterns of the lower temperature phase showed apparent crystalline rings, and this result is reminiscent of the work reported by Iimura and coworkers [37].

One question that was only partially answered for the polymers with this particular mesogenic unit was what length of flexible spacer could be present before the liquid crystallinity was lost. The loss of the mesophase occurred with polymers having more than eight but less than 13 ethylene oxide repeat units; in between these the polymers started to take on the properties of PEO itself. The inability to form a mesophase is apparently due to some type of dilution effect that decreases the thermal stability of the LC domains.

3.3 Polysiloxane Spacers

Polysiloxanes are known to have very low glass transition temperatures and are considered to be much more flexible than polymers of the spacer types discussed above. Quite recently, several examples of polymers using polysiloxane flexible spacers have been synthesized and were characterized by conventional methods.

We prepared polymers containing the mesogenic structure 25 of Table 1, both in the homopolymer form and as copolymers containing polysiloxane and decamethylene flexible spacers, and these polymers were compared in their properties to the earlier polymers prepared with a polymethylene spacer only [56]. The polymer with mesogenic structure 32 and the siloxane spacer showed a low molar enthalpy for the melt transition, while the copolymer and the polymer with the polymethylene spacer showed quite high melt enthalpies.

The range of mesophase stability and melting temperature was greater for the polymer with a polymethylene spacer than that with a siloxane spacer; however, the size of the entropy of clearing was reversed. In all such measurements, the properties of the copolymer fell between these two extremes, except for the case of the range of mesophase stability, which was greater for the copolymer than the two homopolymers. In all cases, the mesophases were identified as nematic from characterization by polarized-light microscopy.

Another homopolymer of structure 32 showed the same trends, except that the range of mesophase stability of its polymethylene copolymers was intermediate between those of the homopolymers, and the copolymer had the highest heat of melting. The order of the entropies of clearing was also reversed, with the polymethylene homopolymer having the highest entropy of clearing. Again the mesophase was concluded to be nematic from their Schlieren textures on the polarizing microscope.

Another series of polymers containing polysiloxane flexible spacers were prepared by Ringsdorf and coworkers, and these are represented by mesogenic structures 22 and 28 to 30 in Table 1 [39]. Unlike polymers containing polymethylene spacers which exhibited melt transitions, many of these polymers showed only a glass transition, which was generally well below 0 °C, and below this temperature the polymers still possessed a mesophase morphology. These results suggest the existence of a glassy liquid crystalline state.

As a result of their low T_g values and lack of crystallinity, many of these polymers showed liquid crystallinity at room temperature. The liquid crystalline mesophases of the monomers were identified as nematic [40], but the polymeric mesophases were not identified, although they possessed a very broad thermal stability between their T_g and their clearing transitions. A mesophase temperature stability of up to 170 °C was observed for the polymers with bicyclohexane central mesogenic units. These polymers showed decreases in T_g and T_i with increased spacer lengths.

The lower transition temperature of the siloxane-containing LC polymers compared to the polymers with polymethylene and poly(alkylene oxide) spacers can be attributed to the greater flexibility, bulky structure and irregular conformations of the siloxane units in these polymer chains. These effects could be expected to reduce the effective chain-packing and interchain forces and, in this way, reduce the transition temperatures and possibly the degree of liquid crystalline order. It appears in the case of the polymers prepared by us, that the siloxane spacer reduced either or both the interchain interactions and the degree of crystallinity of the polyester. The Ringsdorf polymers showed a complete loss in the degree of crystalline order.

To conclude this discussion, it is quite obvious that the flexible spacer plays a very important role in determining not only the transition temperature but also the type of mesophase. In almost all cases, an increase in flexible spacer length leads to a decrease in transition temperatures. For the polymethylene spacers, this decrease is associated with an even-odd effect, with lower transition temperatures for the odd-numbered members of the series. For some mesogenic types, this effect can also be accompanied by a change from nematic to smectic mesophases.

Polymers with poly(ethylene oxide) spacers were shown to have transition temperatures similar to those for polymers with polymethylene spacers of the same length, but the mesophases formed could be quite different, which again emphasizes the importance of the spacer in determining mesophase morphology. Furthermore, polymers with polysiloxane spacers and mesogenic units, identical to those present in polymers with polymethylene spacers having a comparable number of bonds, showed different mesophase structures in addition to much lower transition temperatures.

All of these observations strongly indicate that at least part, if not all, of the spacer is an integral part of the total mesogenic structure in the sense of the use of the term to describe the groups that are responsible for causing the liquid crystal organization. In any case, it is quite apparent that the main chain spacers serve for more than simply providing freedom of motion and assembly for the mesogenic groups, which is considered to be their principal function in side chain LC polymers.

3.4 Effect of Substituents in the Spacer

The addition of substituents to the flexible spacer could be expected to influence the properties of mesophases formed by LC polymers, and the presence of methyl and ethyl pendent groups has been found to lower transition temperatures in several series of polymers. The ethyl group lowered transition temperatures approximately twice as effectively as the methyl group [22]. When a methyl group is added to the units in poly(ethylene oxide) spacers, that is, in the use of poly(propylene oxide) spacers, the resulting LC polymers have greatly reduced melting temperatures (or are non-crystalline), but only slight changes in the clearing temperatures occur [101]. These observations can be explained by the hindered packing in the solid state, caused by the asymmetrically positioned spacer pendent group, which does not affect the mesophase packing because of the greater interchain distances in the LC melt state.

It is also known that in side-chain LC polymers the copolymerization of optically active monomers with mesogenic monomers, in the same manner as the mixing of optically active compounds with nematic low molecular weight compounds, can induce the formation of a cholesteric mesophase. Therefore, it is expected that inclusion of chiral spacers in main chain liquid crystal polymers, which would be nematic in the absence of the chiral centers, will lead to cholesteric products. However, only a small number of reports have appeared to date which describe the preparation and properties of such polymers.

Strzelecki and coworkers [45] found that the use of (+) 3-methyladipic acid in the synthesis of liquid crystal polyesters led to a cholesteric polymer. They prepared a homopolymer and a series of copolyesters of the following structures:

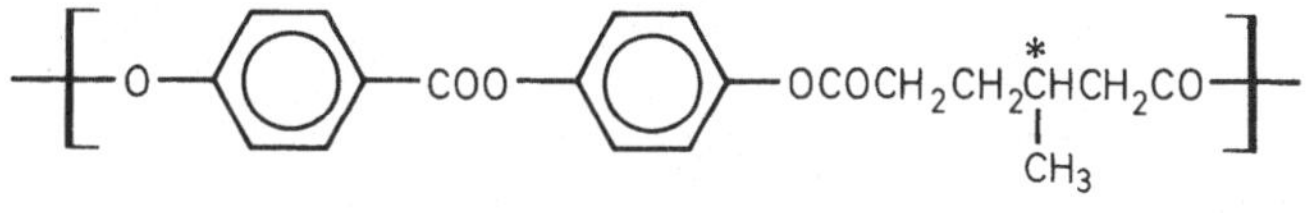

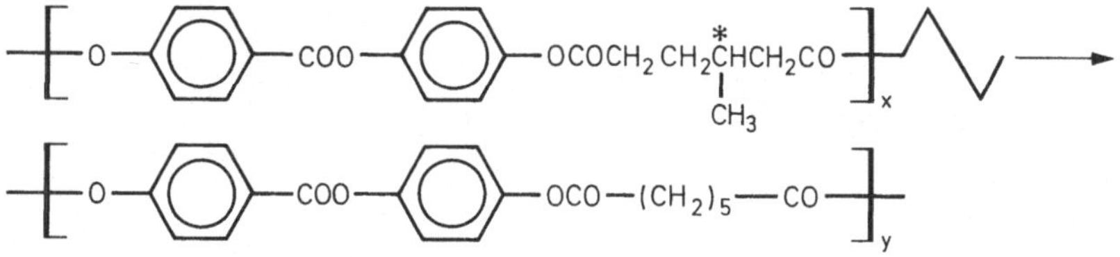

The wavelength of iridescent light reflected by the copolymer melt increased as the content of the non-chiral spacers increased, indicating that the pitch spacing of the helical packing of the nematic layers increased accordingly. That is, this observation can be explained by the assumption that the molecular twist angle increased in a simple relationship to the amount of chiral spacers in the copolymer chain leading to a shorter pitch.

Blumstein and coworkers [46, 47] studied the cholesteric behavior of polyesters with azoxybenzene mesogenic units and the same chiral spacer, (+) 3-methyladipic acid. They could clearly observe oily streak textures, which are typical of low molecular weight cholesterics, for the following homopolymer and copolymers:

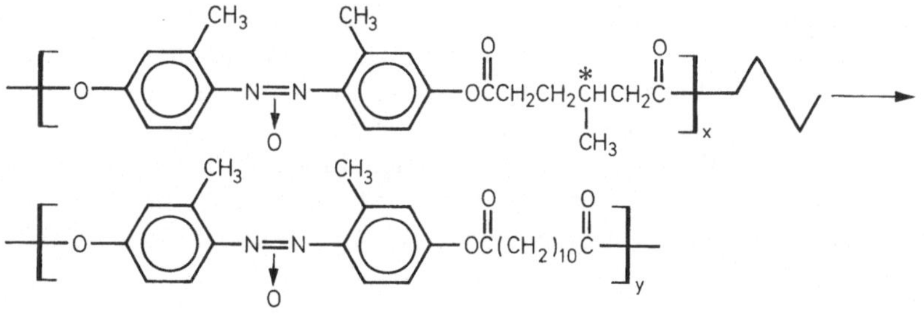

In contrast, a racemic mixture of the 3-methyladipic acid residues in the polyester resulted in the formation of a nematic phase. Similar cholesteric polyesters based on azobenzene mesogenic units and (+) 3-methyladipic acid were reported by Japanese workers [48].

Recently Krigbaum and coworkers [49] prepared cholesteric polyesters from the reaction of 4,4'-dihydroxy-α-methylstilbene with mixtures of (+) 3-methyladipic acid and adipic acid. They could change the morphology of the cholesteric state formed by the (+) 3-methylapidic acid homopolymer by mixing it either with a low molecular weight nematic compound or with a nematic polyester made from adipic acid. The 50:50 copolyester showed a predominately planar texture, which could be retained in the solid state by quenching. The quenched film maintained a deep blue color at room temperature.

A series of cholesteric polyesters were also prepared in our laboratory [50] from either (+) 3-methyladipic acid ot (+) 3-methyl-1,6-hexanediol with the following structures:

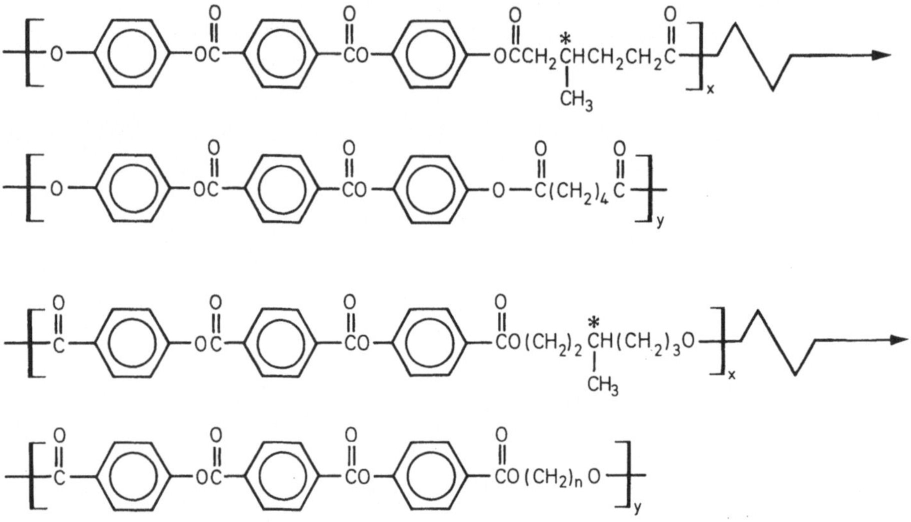

Cholesteric oily streak textures could again be clearly observed for these LC polymers containing chiral spacers. In general, the pitch of the helical packing increased in a regular manner not only with temperature, as judged by an iridescent color, but also with the amount of achiral component in the copolymers.

Unfortunately, there is no report on the detailed physical characterization of these polymers. Such information as unidirectional twist angle and form optical rotation, as well as their dependence on chemical structures and temperature, can be very useful in further understànding the molecular orientations of the polymers in the cholesteric phase. In contrast, a number of studies have been made on the physical-chemical properties of cholesteric lyotropic polymer systems, especially polypeptides.

4 Main Chain LC Copolymers

Depending on their detailed structure, copolymers can have very different properties from the average properties of the corresponding homopolymers. Several reports have appeared on the study of structure-property relationships of liquid crystalline copolymers having mesogenic units and spacers in the main chain.

Roviello and Sirigu [23] prepared copolymers with a single mesogenic group by varying the ratio of two spacers of different length, as follows:

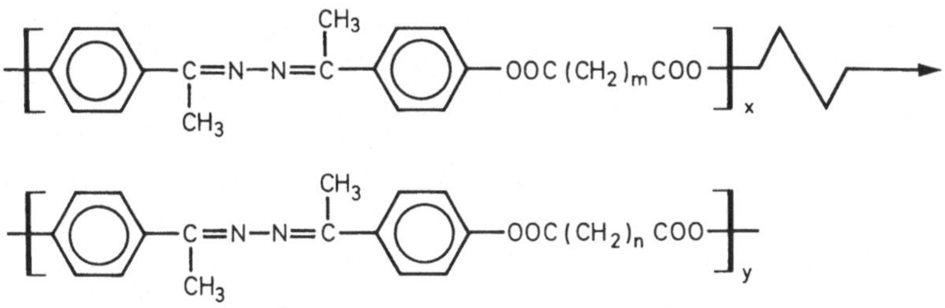

where m and n were 6 and 10, respectively. Clearing temperatures and ΔS_i changed regularly with composition. The temperature ranges of the mesophases of the co-polymers were wider than those of the homopolymers. The maximum mesophase temperature range was observed for the 50/50 copolymer. All of the copolymers were nematic.

A few series of azo and azoxy group containing liquid crystalline copolyesters were prepared by Iimura and coworkers [19], and their phase transition temperatures were examined. No unusual phenomena were found, although monotropic mesophases were observed for the following copolyesters, depending on the combinations and fractions of alkylene groups:

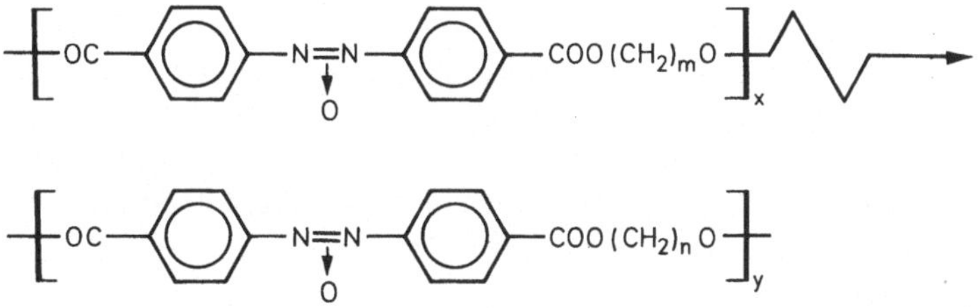

In general they also found that the copolyesters tended to have wider temperature ranges of mesophase stability, in accord with Roviello and Sirigu's observations.

The copolyesters of Griffin and Havens [13], which were discussed earlier in the section on the odd-even effects in LC polymers, also showed similar smooth changes in T_m and T_i. However, close examination of the data given for transition temperatures of the copolyesters reveals that odd-odd combinations of the number of methylene units of the flexible spacer tend to result in wider temperature ranges of mesophase stability than either odd-even or even-even combinations. We [87] have also observed the same trend in the following series of copolyesters:

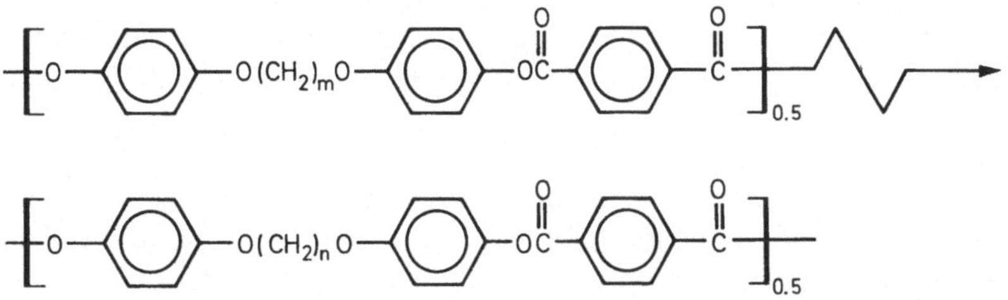

For example, the copolyester containing the combination of 7 and 9 in m and n, respectively, showed a nematic mesophase in the temperature range from 166 to 277 °C, while the mesophase temperature range for 5/10 and 8/10 combinations was only 41 and 89 °C, respectively. The wider mesophase temperature range observed for odd-odd combinations was not due to the intrinsic thermal stability of the mesophase of the copolymer, but rather it was due to the greater degree of suppression in melting temperature of the resulting polymers.

Recently Blumstein and coworkers [47, 86] reported on the thermotropic properties of a series of main chain copolyesters with different azoxybenzene mesogenic units and flexible spacers consisting of varying ratios of (+) 3-methyl adipic acid and 1,12-dodecanedioic acid. Melting temperature of the copolyesters showed minimum values for either the 50/50 compositions or the 25/75 combination of the two spacer components, depending on the nature of the mesogenic units. However, T_i tended to decrease linearly with the content of the 1,12-dodecanedioic acid unit, except in one case. They also observed that the cholesteric pitch of the copolyesters seemed to increase as the concentration of the achiral unit, dodecanedioic acid, increased. Similar observations were reported earlier by Strzelecki's group [41].

The thermodynamic properties of the phase transitions of LC copolymers are not yet well documented, even though odd-odd combinations in the number of methylene units of the spacers seem to have lower ΔS_i values, indicating a less ordered structure. Copolymers of mixed mesogenic units also have not been sufficiently studied to make an attempt to analyze the data [87].

Direct comparison of the thermal behaviors of copolymers requires very careful analysis of the experimental data because the sequence distributions and molecular weights, in addition to compositional variations, should be accounted for. However, it may be safe to conclude that copolymers possess wider temperature ranges of mesophase stability, but give rise to less ordered mesophases.

5 Characterization of LC Polymers

5.1 Microscopy

The characterization of liquid crystals by polarized light microscopy is the most straightforward method available and, whenever possible, it should be carried out in the initial stages of an investigation on new polymers. Thermal analyses alone can be misleading. In this procedure, a thin layer of the melt is kept at constant temperature on a hot-stage and observed between crossed polars. The appearance or "texture" of the melt is dependent on the structure of the mesophase, and, therefore, it is often possible to directly identify the type of mesophase present by this method. A good review of the microscopy of liquid crystals appears in the books by Hartshorne [51] and by Demus and Richter [52].

Texture observation of low molecular weight compounds is usually much easier than that of their polymeric counterparts. Lower melt viscosity presumably results in the quicker appearance of the identifiable melt structures of low molecular weight

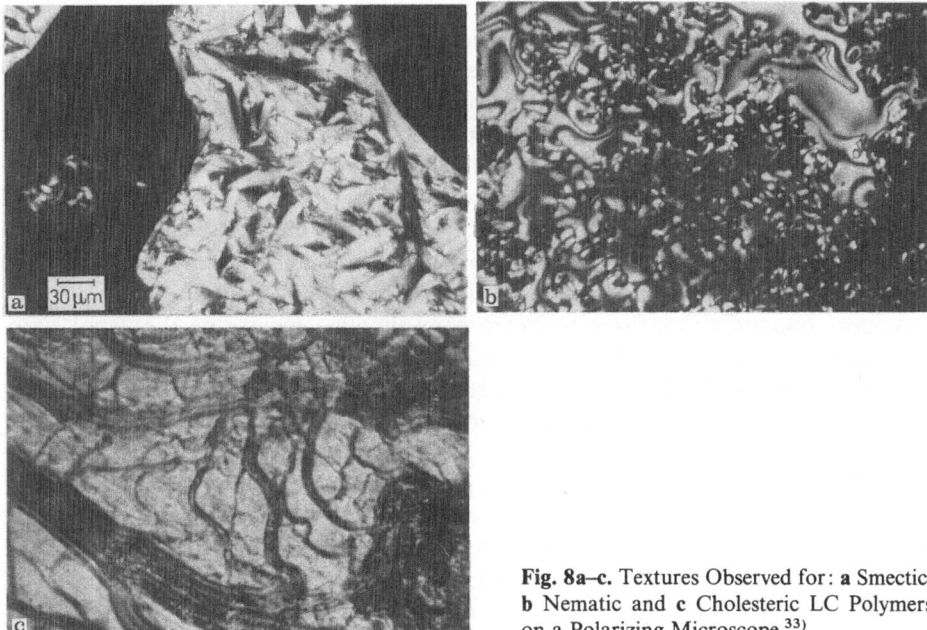

Fig. 8a–c. Textures Observed for: **a** Smectic, **b** Nematic and **c** Cholesteric LC Polymers on a Polarizing Microscope [33]

compounds. Polymeric materials on the other hand sometimes take minutes or hours to show recognizable textures, and in that time period the polymers can decompose. In fact several authors have described polymers which showed no definite texture, possibly because of their high molecular weight, and consequently were usually identified as nematic [12, 18, 28]. In general, however, the textures exhibited by polymers are identical to those of low molecular weight compounds. The threaded Schlieren texture, typical of nematics, and the focal-conic or fan-shaped textures typical of smectics, are the ones most often cited in the literature. Cholesteric polymers, like their low molecular weight counterparts, also produce observations of "oily streaks" and parallel disclinations. Examples of typical polymeric textures are shown in Fig. 8.

Among the more difficult textures to identify is the homeotropic or pseudo-isotropic texture [35, 51, 52]. The mesophase appears black or isotropic when observed between crossed-polars because the director of the mesogenic unit is oriented perpendicular to the surface of the glass slide and parallel to the light beam. The perpendicular orientation of the mesogenic unit and, therefore, of the polymer molecules is presumably due to an interaction between the polymer and the glass surface, and the formation of a homeotropic texture can be promoted by the use of very clean surfaces.

A method of avoiding the formation of the homeotropic texture involves the use of specially treated glass surfaces. By obliquely evaporating SiO_2 on the glass surface, the new surface causes an alignment of the mesophase at an angle to the incident light [36]. A homogeneous texture can also be formed by surface treatments, and in this state the mesophase is more easily identified. Rubbing of the glass surface can also produce similar effects [51].

The mesophase exhibiting a homeotropic texture can still show stir opalescence, which is a method of identification particularly suited for thermotropic polymers. This somewhat crude method of characterizing a liquid crystalline material is performed by shearing a thin film of the mesophase and looking for momentary appearances of turbidity in the otherwise transparent melt. No microscope is needed to observe stir opalescence, but simple shearing of the homeotropic melt between crossed-polars can also reveal the mesophase.

Identification of mesophases by microscopy is subject to a great deal of subjective interpretation. It is advisable, therefore, to use this method in conjunction with other types of characterization, such as X-ray diffraction or thermal analysis.

A more quantified approach to microscopy involves measuring the depolarized light intensity with a photocell, and plotting this quantity as a function of temperature. This approach is particularly useful in determining the clearing transition when rather broad transitions, quite usual for polymers, are obtained [34]. Unless care is taken to ensure a constant field of view and constant sample thickness, this method should be considered to be only semiquantitative as a measure of depolarized light intensity.

Another technique which uses microscopy is based on the miscibility of compounds with identical mesophases and was developed by the Halle liquid crystal group for model liquid crystals. Noel has applied this method to mixtures composed of well-known model liquid crystals with polymeric liquid crystals [35, 49]. Assuming that the method is applicable to mixtures of polymers and low molecular weight compounds, the type of mesophase can be positively identified if the polymer and model are miscible.

Transition temperatures and the number of transitions can be determined by thermal analysis as performed by Griffin and Havens [28]. By combining the thermal observations with the microscopy observations, a phase diagram can be drawn which indicates whether or not the phases are miscible. An example of this analysis is given in Fig. 9, which shows the phase diagram of a mixture of Polymer 18 of Table 1 with

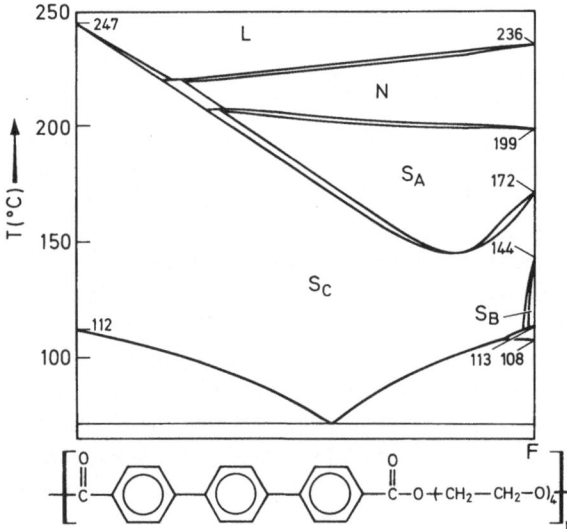

Fig. 9. Phase Diagram for a Physical Mixture of Polymer 18 and a Model Compound [35]

the model compound shown below that Noel used to confirm the smectic C phase of the polymer:

$$C_4H_9 - \bigcirc - N{=}CH - \bigcirc - CH{=}N - \bigcirc - C_4H_9$$

The addition of chiral compounds to either nematic or smectic C phases can be used to convert the mesophases to the cholesteric or twisted smectic C phases, respectively [35]. In this way the possibility of other mesophases being present can be eliminated, and these two mesophases can be differentiated by X-ray diffraction.

The highly viscous nature of polymeric mesophases could possibly prevent the mixing of the mesophases of polymers and certain model compounds. Therefore, while the miscibility of model compounds with a polymer may be used to identify the type of mesophase, the lack of compatibility does not necessarily suggest that the mesophases are not the same. That is, some model compounds and polymers with the same mesophase may even be inherently incompatible. As a result, some judgment is required in order to make both the proper choice of the model compound and the correct assignment of the mesophase when using this technique.

5.2 X-Ray Diffraction

The study of liquid crystalline materials by X-ray diffraction has until recently been confined to low molecular weight compounds because of the lack of availability of suitable polymeric mesogens. Based on the observations by X-ray diffraction made with small molecules as reviewed by Azaroff [89] a system for identifying the type of mesophase has been developed by DeVries [88], and this system is now being applied to polymeric liquid crystals.

Studies of thermotropic polymers by wide-angle X-ray diffraction, WAXD, in the liquid state may be very useful in the identification of the structure of the liquid crystalline melt, because microscopy experiments cannot always be counted on to produce an unambiguous determination of the type of mesophase present. In WAX diffraction patterns, nematic structures produce a diffuse ring (4—5 Å) on a flat film diffractogram from the interchain spacings, which can also form rings at smaller diffraction angles. These rings are caused by the "cybotactic" structures [90], in which some of the mesogenic repeat units are organized into more highly ordered domains in the melt in comparison to the average nematic domain.

Cholesteric mesophases in general resemble the nematic mesophase when observed by both small-angle X-ray diffraction, SAXD, and WAXD. The smectic mesophase, in contrast, produces both diffuse rings at 4—5 Å and sharp rings at a distance generally, but not necessarily, equal to the repeat length of the monomer unit, between 15 and 50 Å.

When used in conjunction with microscope observations, the X-ray diffraction spectra, examples of which are shown in Fig. 10, can usually be used to accurately identify the mesophase. Oriented and quenched samples can also provide information about these phases. Because there are several types of smectic mesophases, one must

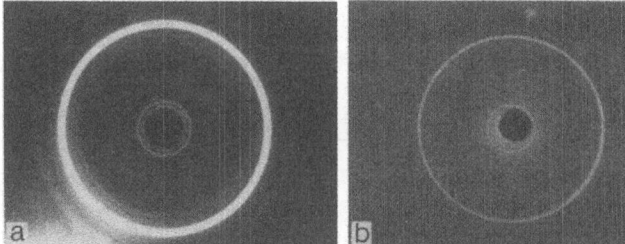

Fig. 10a and b. X-ray Diffraction Patterns of Thermotropic Polymers: Polymer 15 Obtained at Room Temperature, **a** and in the Liquid Crystal State, **b** [73]

Temperature (°C)	Spacings (Å)
25 (annealed, sample b)	3.47 (w); 3.92 (w); 4.25 (s); 4.69 (m); 4.96 (m); 7.60 (w); 9.78 (w); 15.8 (m); 19.2 (s)
170 (sample a):	4.33 (s); 15.8 (w)
(sample b):	4.35 (s); 14.6 (w); 19.5 (s)
220 (sample a):	4.46 (s); 16.4 (w)
194 (sample b):	4.46 (s); 17.0 (s)

orient the direction of alignment of the mesogenic units with respect to the direction of orientation of the sample to get an indication of the mesophase structure. Alignment in the sample can be caused by shearing, electric fields or magnetic fields. Additional information can be gained from the smectic E mesophase, which is highly structured, and therefore, the diffraction pattern of its melt shows many rings and resembles that of a solid.

Despite its usefulness, X-ray studies of the melt cannot always be made. Many polymeric samples have relatively high melting points at temperatures where the polymers can degrade. These temperatures are often above the operating range of the sample heaters, and in any case, the extended heating of polymer samples at such temperatures can cause degradation. Quenching of the mesophase is useful if crystallization does not occur or if the crystal size is so small that it does not interfere with mesophase diffraction because the presence of a crystalline pattern in addition to liquid crystal diffraction pattern can greatly confuse the interpretation of the latter. Many of the LC polymers have been studied by X-ray diffraction both in the melt and in oriented and unoriented solids. These polymers have been generally identified as nematic because only the interchain spacings were observed. In oriented samples an interchain spacing was often identified and ascribed to a cybotactic structure.

Strzelecki has studied the orientation of a nematic polymer in its LC state in a 0.3 T magnetic field by X-ray diffraction and found that the sample had an order parameter of 0.64 [61]. The orientation of LC polymers in a magnetic field has also been observed by others using X-ray techniques [47]. Extruded fibers have been studied by Roviello and Sirigu to measure the degree of orientation achievable in mesomorphic polymers by shear field, but only solid samples were studied and even though they were oriented, the results gave little quantitative indication of the order of the mesophase [24].

A very interesting area of investigation, where X-ray diffraction coud be useful, would be in the measurement of spacer extension in the liquid crystalline melt. The study of polymers forming the smectic mesophase would be the most informative, since a measure of the interlayer spacing should answer the question of whether or not the spacer is fully extended in the mesophase. To date, however, only a few polymers forming smectic mesophases have been studied by X-ray diffraction.

Skoulios [56] and Blumstein and coworkers [20] have studied polymers that were thought to form a tilted smectic B mesophase. In the case of Polymer 40 of Table 1, a spacing of 25 Å was observed in addition to one at 4.5 Å. Another polymer of similar structure, Polymer 41, was also observed to be smectic B, but no further details of the X-ray pattern were given. The smectic polymers, shown as structures 8 and 15 in Table 1, had a spacing at 4–4.5 Å and another at 15–20 Å [19]. Noel reported that X-ray studies confirmed the presence of a smectic mesophase in polymer 17, but no other data were presented [36]. A monotropic smectic phase was observed by Roviello and Sirigu [59] with mesogenic structures 2 and 5 of Table 1 and a diethylene glycol spacer. In addition to the expected halo, Polymers 2 and 5 exhibited sharp rings at 21.8 and 20.6 Å, respectively.

Other polymers with oligo(ethylene oxide) spacers have been observed. The polymers of Iimura and coworkers, with mesogenic structures 10 and 12 of Table 1 and spacers of di-, tri- and tetraethylene oxide, had X-ray patterns in the melt that were virtually identical to those of the solid [37]. This result was also observed for Polymer 37 with a tetraethylene oxide spacer prepared by us [62]. Such patterns are consistent with a smectic E mesophase which maintains much of the solid state order in the liquid crystalline melt.

Recently we studied several polymers of structure 36 in Table 1 that were identified by texture observation as being smectic [32]. X-ray diffraction patterns showed a spacing at approximately 5 Å and others at 28, 31, and 29 Å for the polymers with 9, 10, and 12 methylene groups in the spacer, respectively. Such repeat lengths are not consistent with the presence of fully extended chains in the smectic A mesophase for these polymers. Instead, the polymer with a nonamethylene spacer appears to form a smectic C phase in which the polymer chain is tilted with respect to the layers, as suggested by the small interlayer distances.

In the polymers of Roviello and Sirigu discussed above, the chain direction was perpendicular to the layers, yet the fully extended repeat unit of the polymer with mesogenic structure 2 had a length of approximately 29 Å while a spacing of only 20.6 Å was observed [59]. These examples indicate that the spacer may not always be completely extended in the liquid crystalline state, but the spacer is certainly much more extended than if it were in a random coil configuration.

A large number of low molecular weight smectogens have been prepared and studied by X-ray diffraction methods [91]. At this time relatively few smectic polymers have been prepared and identified, and the smectic D, G, and H phases have not, as yet, been reported in polymers.

5.3. Thermal Analysis

In addition to characterization of the mesophase by microscopy and X-ray diffraction, the most common technique for studying liquid crystalline compounds is by thermal analysis. This technique both reveals transition temperatures and gives a measure of the heat of transition. A large amount of literature exists concerning the thermal behavior of low molecular weight compounds, and most studies of polymeric and model LC compounds have been performed using a DSC or DTA instrument.

The melting temperatures of polymeric materials can be greatly affected by the sample history, but it is now commonly agreed that the clearing transition is much less subject to effects of thermal treatment. If it can be assumed that the clearing transition is an equilibrium process, then the entropy of this transition, ΔS_i, can be calculated from the clearing temperature and from the enthalpy of the transition. Values of ΔS_i give some indication of the order present in the system when the iso-tropic state is assumed to have equal disorder in all systems.

In low molecular weight liquid crystals the enthalpy of clearing is usually a small fraction, approximately 3–5%, of that of melting [54]. In polymers, the lack of complete crystallinity in the solid, as well as effects of melt history and the nature of the polymer mesophase, can cause the values of enthalpy of the melting and clearing transitions to be very similar.

As shown in Fig. 11, the melting peak obtained by DSC is often structured and the clearing transition can be rather broad. The structuring of the melting peak may be an indication of polymorphism in the solid state [42], but more often it is caused by successive meltings and recrystallizations of regions of imperfect crystallinity. The existence of more than one mesophase in the melt has also been observed, and this behavior leads to the presence of two or more liquid crystal transitions in the thermo-gram [34].

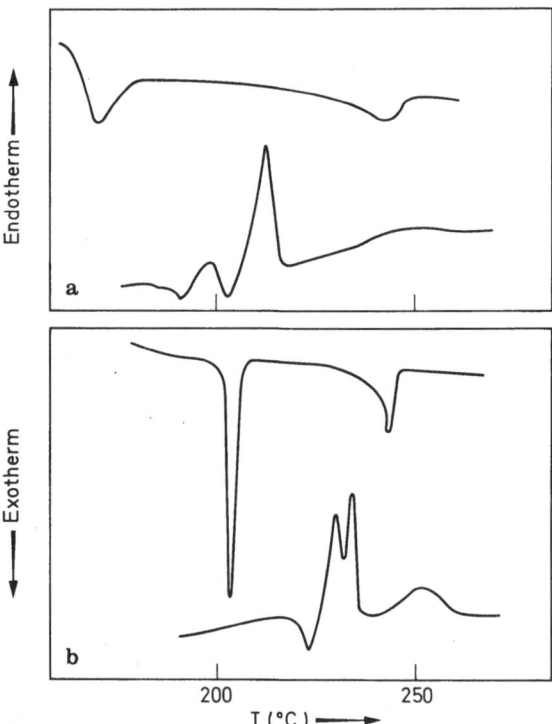

Fig. 11 a and b. DSC Thermogram of Polymer 36 with (a) a Nonamethylene Spacer and (b) a Dodecamethylene Spacer [33]

Table 4. Effect of Mesogenic Group and Flexible Spacer on Entropy of Clearing, ΔS_i, of LC Polyesters

Mesogenic Unit	n in Polymethylene Spacer	LC Type[a]	ΔS_i cal mol^{-1} K^{-1}[b]
1)	6–10, 12	N or S	1.5–3.5
2)	7–10	N	0.5–1.5
3)	2–10	N and/or S	1.0–5.4
4)	2, 4, 6	S	3.0–5.5
5)	8–14	N	2.0–4.5
6)	8–14	N	1.5–4.0
7)	2—14	N	0.5–4.5
8)	6, 8, 10	N	1.5–4.5

[a] N — nematic; S — smectic;
[b] molecular weight of average repeat unit used to calculate number of moles repeat unit

Listed in Table 4 are the values of ΔS_i for some of the polymers studied for this property. Only polymers with polymethylene spacers are listed in this table, and where possible, the type of mesophase formed is also given. A remarkable feature of most of these examples is that, when homologous series were studied, an even-odd effect in the entropy of clearing was observed which was quite similar to the even-odd properties of both the melting and clearing transition temperatures. It should be noted that the values of ΔS_i for model compounds are generally much lower than those for polymers. This observation may suggest that the order present in a polymeric mesophase is much higher than is found in a model system.

Considering the number of possible defects in a polymer mesophase from chain entanglements, or from the chains connecting the ordered regions, the possibility of a higher order in the polymer mesophase is surprising. Another explanation, however, might be that the higher values of ΔS_i result from mixing between the spacer and mesogenic groups at high temperatures. This suggestion could qualitatively explain why polymers with the more polar poly(ethylene oxide) spacers have lower ΔS_i values than those with polymethylene spacers, for most polar mesogenic units, but this situation is not true for polymers with non-polar terphenyl mesogenic groups.

The nematic state is less highly organized than the smectic state, and this difference would be expected to be reflected in similar differences in the entropy of clearing. In fact, as seen in Table 4, values of ΔS_i are quite independent of the type of mesophase present, and several researchers have reported on series of polymers in which a nematic phase had a value of ΔS_i higher than that of a smectic phase [22]. However, it must be cautioned that this apparent discrepancy could be caused by molecular weight differences between these polymers, because in the low molecular weight range often encountered with polyesters, polymers with higher molecular weights can have higher values of ΔS_i regardless of the type of mesophase present.

As briefly mentioned earlier, thermal studies have been used in conjunction with characterization by polarized light microscopy to determine the miscibility of polymeric and small molecule liquid crystals [28], and low molecular weight mesogens, of the same or different types of liquid crystallinity, can also be used as plasticizers or diluents for polymers, as demonstrated in a study involving side chain liquid crystalline polymers [102].

The thermal and thermodynamic properties of LC polymers make it clear that the flexible spacer has a great influence on both the transition temperatures and transition entropies. This conclusion is again evidence of the vital part played by the flexible spacer in determining the organization and degree of order in the mesophase, and it also points out the critical interplay which occurs between the mesogenic unit and the flexible spacer.

5.4 Rheological Properties

The rheology of low molecular weight thermotropic compounds has been a subject of considerable theoretical and experimental analysis [74-79]. In general, liquid crystals are easily oriented by surfaces, electromagnetic fields and mechanical stress or shear, and the degree of orientation, in turn, affects their melt viscosity. The rheological behavior of a liquid crystal is known to be greatly dependent on the nature and also on the texture of its mesophase.

Among the many different classes of thermotropic polymers, only a limited number of polyesters based on aromatic ester type mesogenic units have been studied by rheological methods, beginning with the publication by Jackson and Kuhfuss [8] of their work on the p-oxybenzoate modified polyethylene terephthalate, PET, copolymers. They prepared a series of copolyesters of p-hydroxybenzoic acid, HBA, and PET and measured the apparent melt viscosity of the copolymers as a function of their composition by use of a capillary rheometer. On inclusion of low levels of HBA into PET, the melt viscosity increased because of partial replacement of the more flexible ethylene glycol moiety by the rigid HBA units. However, for approximately constant molecular weights, the melt viscosity passed through a maximum value and then decreased when the fraction of HBA units exceeded 30 mole %. This composition coincided with the appearance of liquid crystalline anisotropy in the melts. The melt viscosity reached a minimum at approximately 60 mole % of HBA units as shown in Fig. 12. The melt viscosity-composition behavior of thermotropic copolymers is conceptually the same as the solution viscosity-concentration behavior of lyotropic polymers [80].

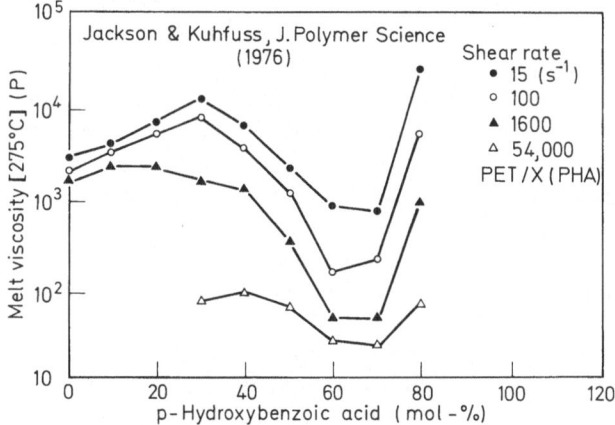

Fig. 12. Viscosity Versus Mesogenic Comonomer Concentration of Copolymers of Ethylene Terephthalate and p-Hydroxybenzoic Acid (after Jackson and Kuhfuss) [8]

Jerman and Baird [76] recently conducted rheological studies on the same copolymers using an Instron capillary rheometer. They also measured die swell and entrance pressures. They observed that the viscosity of the 60 mole % HBA/PET copolymer was two orders of magnitude lower than that of PET when compared at the same temperature of 285 °C, which is similar to the results reported earlier by Jackson and Kuhfuss [8]. Die swell of the copolymers was highly temperature dependent. In general, it increased with temperature, while for nonliquid crystalline isotropic polymer melts the opposite is normally observed. Another anomalous phenomenon observed was that, although the die swell data indicated that negligible elastic recovery occurred, the values of the ratio of entrance pressure loss to wall shear stress were much higher for liquid crystalline melts than for PET. The bases for such anomalies are not yet clear.

Wissbrun [82] earlier observed a very long relaxation time and high elasticity for anisotropic melts of aromatic polyesters, as well as several other types of flow anomalies. Unfortunately, in most of these earlier studies, the rheological behavior of liquid crystal melts of polymers could not be directly compared with that of the isotropic phase of the same polymers because of their high clearing temperatures.

More recently Wissbrun and Griffin [83] reported in some detail on the rheological properties of one of the polymers reported earlier by Griffin and Havens [13], Polymer 35 of Table 1 with a decamethylene spacer, in both the nematic and isotropic states. All of the rheological measurements were made on a Rheometrics mechanical spectrometer mostly in the oscillatory shear mode over a frequency range of 1–100 rad/s. For this polymer, the isotropic melt state at 230 to 240 °C showed the usual rheological behavior of other isotropic polymer melts in that the value of the complex viscosity $|\eta^*|$ was almost Newtonian, and $|\eta^*|$ and the storage modulus G' were independent of the direction of frequency change. It was observed that the increase in shear stress and its decay occurred almost instantaneously.

The value of the compliance J' for this polymer was about 10^{-6} cm^2/dyne, which is very close to the values for linear flexible polymers. An activation energy of 11 Kcal deg^{-1} was obtained for viscous flow. This value is fairly close to that of PET, 13.6 Kcal deg^{-1} [84]. In contrast to the isotropic phase, the rheology of the liquid crystalline melt was much more complicated in that the behavior of $|\eta^*|$ was strongly characteristic of shear thinning, and the shape of the flow and the storage modulus curves were upwardly concave. The magnitude of J' and the relaxation times were 1 to 3 orders of magnitude larger than those of the isotropic phase. Nematic phases also showed apparent yield stress behavior. Unlike the behavior of an isotropic melt, the nematic phase did not achieve a steady-state shear stress instantaneously, and its oscillatory flow behavior depended on the shear history of the sample. The modulus and viscosity at lower frequencies depended on the direction of the frequency change and were found to be lower on the descending frequency run than on the ascending one.

The difference in the flow behavior of isotropic and nematic phases seems to be analogous to that observed for low molecular weight liquid crystal compounds. The shear-thinning property of the mesophase was explained by the Leslie-Erickson theory [78, 79] originally proposed for low molecular weight liquid crystals. The shear thinning is said to be the result of a competition between orientation perpendicular to the flow at the wall and a nearly parallel orientation in the bulk of the melt. The apparent stress yield behavior can be rationalized by the presence of domains having different orientations. This representation of the liquid crystal state of thermotropic polymers was proposed earlier by Onogi and Asada [85], and the dependence of the rheological properties on shear history can also be at least qualitatively explained by this mechanism.

Even though the mesophase of the polymers with flexible spacers seemed to display many similar characteristics as those observed earlier for rigid aromatic polyesters, there existed several distinctive differences, which will not be discussed in this review because the subject is covered in detail in a review specifically devoted to the rheology of liquid crystals for rigid rod-like molecules [81]. In that review, the behavior of polymeric liquid crystals was compared with that of low molecular weight compounds, mainly on the basis of the Onogi-Asada model. It is apparent that the study of the rheology of liquid crystal polymers is in its infancy and much more research is needed before one can draw any conclusion based on specific models or theories. Also it should again be cautioned that many of the polymers examined so far were of rather low molecular weights. Needless to say, an understanding of the rheological properties of the liquid crystal phase of polymers is extremely important, not only for theoretical interest but also for the optimization of processing conditions and for the development of maximum physical and mechanical properties of such polymers.

6 Field Effects on LC Polymers

The molecular structures of nematic and cholesteric liquid crystals are highly aniso-tropic and generally show a high degree of polarizability. Such properties can cause many distinctive effects on the application of magnetic or electrical field to LC com-pounds. Because of their anisotropic magnetic susceptibilities and dielectric constants, liquid crystal molecules have a strong tendency to become aligned when placed in a magnetic or an electric field, and their behavior in electro-magnetic fields is well documented. This property is put to use in applications such as display optics. In contrast, the effects of electric or magnetic fields on LC polymer melts have not been studied in much detail.

Krigbaum and coworkers [93] studied electric field induced flow instabilities in the thermotropic HBA/PET copolyesters discussed above [8]. Williams domain patterns [94] could be obtained in the copolymer containing 60 mole % HBA units at high temperatures in a direct current field. Because of the high viscosity and long relaxation times of this polymer in comparison with low molecular weight liquid crystals, considerably longer times (of the order of hours instead of seconds) were required for the complete formation of Williams domains. No Williams domain pattern was observed for the nonliquid crystalline copolyester containing 30 mole % HBA units.

No dynamic scattering mode was observed for the 60 mole % HBA copolymer. However, a pattern of dark, turbulent regions was seen at high fields (150 V and 50 Hz) for the LC copolymers but not for the isotropic copolymer [95,96]. Because this high field pattern for the liquid crystal polymer formed in a minute or less, they proposed that this phenomenon might provide a convenient test for a nematic phase in such polymers. Very thin samples of polymer with liquid crystal compositions formed the variable grating mode [100], but it was assumed that orientation was achieved over only small regions of the samples. It is to be expected that liquid crystal polymers should behave in an electrical field in the same manner as do low molecular weight compounds [97], but because of their higher melt viscosities and longer relaxation times, the response of LC polymers to an electrical field will require longer times and higher fields.

A liquid crystal polyester having azoxybenzene mesogenic units and decamethylene spacers was subjected to a proton NMR study after the polymer was aligned in a mag-netic field of ~ 1 Tesla [98]. Below the melting point of the polymer, the NMR spectrum showed a simple broad line for the mesogenic group, and the peak width decreased with increasing temperature. At the melting point, two additional shoulders appeared, and the spectra became better resolved. The splitting decreased as the temperature increased.

The order parameter in this sample as determined from the NMR spectra varied from 0.88 to 0.72 over the nematic temperature range. This value is higher than the usual values for low molecular weight nematic liquid crystal compounds.

References

1. Kwolek, S. L., Morgan, P. W., Schaefgen, J. R. and Gulrich, L. W.: Macromolecules, *10*, 1390 (1977).
2. Flory, P. J.: Proc. Roy. Soc., *A234*, 73 (1956).

3. Gray, G. W. and Winsor, P. A., ed.: Liquid Crystals and Plastic Crystals, 2 Vol., John Wiley & Sons, N.Y., 1974.
4. Shibaev, V. P. and Platé, N. A.: Polym. Sci. U.S.S.R., *19*, 1065 (1978).
5. Blumstein, A., ed.: "Liquid Crystalline Order in Polymers", Academic Press, N.Y., 1978.
6. Blumstein, A., ed.: "Mesomorphic Order in Polymers and Polymerization in Liquid Crystalline Media", ACS Symp. Series 74, Washington, D.C., 1978.
7. Finkelmann, H., Ringsdorf, H. and Wendorff, J. H.: Makromol. Chem., *179*, 273 (1978).
8. Jackson, W. J. and Kuhfuss, H. F.: J. Polym. Sci.: Polym. Chem. Ed., *14*, 2043 (1976).
9. McFarlane, F. E., Nicely, V. A. and Davis, T. G.: in "Contemporary Topics in Polymer Science", Vol. 2, p. 109, E. M. Pearce and J. R. Schaefgen, eds., N.Y., 1977.
10. Griffin, B. P. and Cox, M. K.: Brit. Polym. J., *12*, 147 (1980).
11. Lenz, R. W., Jin, J.-I. and Feichtinger, K.,: Polymer, *24*, 327 (1983).
12. Strzelecki, L. and van Luyen, D.: Europ. Polym. J., *16*, 299 (1980).
13. Griffin, A. C. and Havens, S. J.: J. Polym. Sci.: Polym. Phys. Ed., *19*, 951 (1981).
14. Antoun, S., Lenz, R. W. and Jin, J.-I.: J. Polym. Sci.: Polym. Chem. Ed., *19*, 1901 (1981).
15. Jin, J.-I., Antoun, S., Ober, C. and Lenz, R. W.: Brit. Polym. J., *12*, 132 (1980).
16. Noel, C., Billard, J. and Fayolle, B.: Preprints Macro IUPAC (Florence), *3*, 286 (1980).
17. Jo, B.-W., Lenz, R. W. and Jin, J.-I.: Makromol. Chem., Rapid Commun., *3*, 23 (1982).
18. Zhou, Q.-F.: Ph.D. Thesis, University of Massachusetts, 1983.
19. Iimura, K., Koide, N. and Ohta, R.: Rep. Prog. Polym. Phys. Jap., *24*, 231 (1981).
20. Blumstein, A., Sivaramakrishnan, K., Clough, S. B. and Blumstein, R. B.: Mol. Cryst. Liq. Cryst., *49* (Letters), 255 (1979).
21. Blumstein, A. and Thomas, O.: Macromolecules, in press.
22. Meurisse, P., Noel, C., Monnerie, L. and Fayolle, B.: Brit. Polym. J., *13*, 55 (1981).
23. Roviello, A. and Sirigu, A.: Europ. Polym. J., *15*, 61 (1979).
24. Roviello, A. and Sirigu, A.: Europ. Polym. J., *15*, 423 (1979).
25. Roviello, A. and Sirigu, A.: Makromol. Chem., *183*, 895 (1982).
26. Krigbaum, W. R., Asrar, J., Tariumi, H., Ciferri, A. and Preston, J.: J. Polym. Sci.: Polym. Letters Ed., *20*, 109 (1982).
27. Roviello, A. and Sirigu, A.: J. Polym. Sci.: Polym. Lett. Ed., *13*, 445 (1975).
28. Griffin, A. C. and Havens, S. J.: J. Polym. Sci.: Polym. Lett. Ed., *18*, 259 (1980).
29. Griffin, A. C. and Havens, S. J.: Mol. Cryst. Liq. Cryst., *49* (Letters), 239 (1979).
30. Van Luyen, D. and Strzelecki, L.: Europ. Polym. J., *16*, 303 (1980).
31. Strzelecki, L. and Liebert, L.: Europ. Polym. J., *12*, 1271 (1981).
32. Ober, C., Jin, J.-I. and Lenz, R. W.: Makromol. Chem., Rapid Commun., *4*, 49 (1983).
33. Ober, C., Jin, J.-I. and Lenz, R. W.: Polymer J. (Japan), *14*, 9 (1982).
34. Galli, G., Chiellini, E., Ober, C. and Lenz, R. W., Makromol. Chem., *183*, 2693 (1982).
35. Fayolle, B., Noel, C. and Billard, J.: J. Phys. (Paris), *40*, C3-485 (1979).
36. Bosio, L., Friedrich, C., Laupretre, F., Noel, C. and Virlet, J.: Preprints, Macro IUPAC (Amherst), 804 (1982).
37. Iimura, K., Koide, N., Ohta, R. and Takeda, M.: Makromol. Chem., *182*, 2563 (1981).
38. Skorokhodov, S. S., Bilibrin, A. Yu., Shepelevsky, A. A. and Frenkel, S. Ya.: Preprints Macro IUPAC (Florence), *2*, 232 (1980).
39. Ringsdorf, H. and Schneller, A.: Brit. Polym. J., *13*, 43 (1981).
40. Aguilera, C., Ringsdorf, H., Schneller, A. and Zentel, R.: Preprints Macro IUPAC (Florence), *3*, 306 (1980).
41. Kleinschuster, J. J., Pletcher, T. C., Schaefgen, J. R. and Luise, R. R.: Ger. Offen, 2,520,819; 2,520,820.
42. Millaud, B., Thiery, A. and Skoulios, A.: Mol. Cryst. Liq. Cryst., *41* (Letters), 263 (1978).
43. Catala, J. M., Majnusz, J. and Lenz, R. W.: Europ. Polym. J., in print.
44. Zhou, Q.-F. and Lenz, R. W.: J. Polymer Sci., in print.
45. van Luyen, D., Liebert, L. and Strzelecki, L.: Europ. Polym. J., *16*, 307 (1980).
46. Blumstein, A. and Vilasagar, S.: Mol. Cryst. Liq. Cryst. Lett., *72*, 1 (1981).
47. Blumstein, A., Vilasagar, S., Ponrathnan, S., Clough, S. B. and Blumstein, R. B.: J. Polym. Sci.: Polym. Phys. Ed., *20*, 877 (1982).
48. Iimura, K., Koide, N., Tsutsumi, Y. and Nakatami, M.: Rep. Prog. Polym. Phys. (Japan), *25*, III-1 (1982).

49. Krigbaum, W. R., Ciferri, A., Asrar, J. and Toriumi, H.: Mol. Cryst. Liq. Cryst., *76*, 79 (1981).
50. Jin, J.-I., Park, H. and Lenz, R. W.: "Synthesis and Properties of New Cholesteric Main Chain Polyesters", presented at "IX. International Conference on Liquid Crystals", Bangalore, India, Dec. 6–10, 1982.
51. Hartshorne, N. H.: "The Microscopy of Liquid Crystals", Microscope Publications Ltd., London, 1974.
52. Demus, D. and Richter, L.: "Textures of Liquid Crystals", Verlag Chemie, Weinherim, 1978.
53. Noel, C. and Billard, J.: Mol. Cryst. Liq. Cryst., *41* (Letters), 269 (1978).
54. Johnson, J. F. and Porter, R. S., ed.: "Liquid Crystals and Ordered Fluids", Plenum Press, N.Y., 1974.
55. Ober, C., Lenz, R. W., Galli, G. and Chiellini, E.: Macromolecules, *16*, 1034 (1983).
56. Millaud, B., Thiery, T., Strazielle, C. and Skoulios, A.: Mol. Cryst. Liq. Cryst., *49* (Letters), 299 (1979).
57. Guillon, D. and Skoulios, A.: Mol. Cryst. Liq. Cryst., *49* (Letters), 119 (1978).
58. Roviello, A. and Sirigu, A.: Gazz. Chim. Ital., *110*, 403 (1980).
59. Maier, W. and Saupe, A.: Z. Naturforschg., *149*, 882 (1959).
60. Jo, B.-W., Jin, J.-I. and Lenz, R. W.: Europ. Polym. J., *18*, 233 (1982).
61. Liebert, L., Strzelecki, L., van Luyen, D. and Levelut, A. M.: Europ. Polym. J., *17*, 71 (1981).
62. Azaroff, L. V.: U. of Connecticut, private communication.
63. Roviello, A. and Sirigu, A.: Makromol. Chem., *180*, 2543 (1979).
64. Roviello, A. and Sirigu, A.: Makromol. Chem., *181*, 1799 (1980).
65. de Jen, W. H., van der Veen, J. and Goosens, W. J. P.: Solid State Commun., *12*, 405 (1973).
66. Stenschke, H.: Solid State Commun., *10*, 653 (1972).
67. Gray, G. W. and Mosely, A.: J. Chem. Soc., Perkin Trans., *2*, 97 (1976).
68. Gray, G. W.: J. Phys. (Paris), *36*, 337 (1975).
69. Gray, G. W. and Harrisen, K. J.: Symp. Faraday Soc., *5*, 54 (1971).
70. Pines, A., Ruben, D. J. and Allison, S.: Phys. Rev. Letters, *33*, 1002 (1974).
71. Marcelja, S.: J. Chem. Phys., *60*, 3599 (1974).
72. Pink, D. A.: J. Chem. Phys., *63*, 2533 (1975).
73. Blumstein, A., Sivaramakrishnan, K. N., Blumstein, R. B. and Clough, S. B.: Polymer, *23*, 47 (1982).
74. Leslie, F. M.: "Theory of Flow Phenomena in Liquid Crystals", in Advances in Liquid Crystals, Vol. 4, Brown, G. H., ed., Academic Press, N.Y., 1979.
75. de Gennes, P. G.: "The Physics of Liquid Crystals", Clarendan Press, Oxford, 1974.
76. Ericksen, J. L.: "The Mechanics of Nematic Liquid Crystals", in "The Mechanics of Visco-elastic Fluids", Rivlin, R. S., ed. (AMD Vol. 22), ASME, N.Y., 1977.
77. Porter, R. S. and Johnson, J. F.: in "Rheology, Vol. 4", Eirich, F. R., ed., Academic Press, N.Y., 1967.
78. Fisher, J. and Fredrickson, A. G.: Mol. Cryst. Liq. Cryst., *8*, 267 (1969).
79. Tseng, H. C., Silver, D. L. and Finlayson, B. A.: Phys. Fluids, *15*, 1213 (1976).
80. Jerman, R. F. and Baird, D. G.: J. Rheol., *25*(2), 275 (1981).
81. Wissbrun, K. F.: J. Rheol., *25*, 619 (1981).
82. Wissbrun, K. F.: Brit. Polym. J., *12*, 163 (1980).
83. Wissbrun, K. F. and Griffin, A. C.: J. Polym. Sci.: Polym. Phys. Ed., *20* (1982).
84. Gregory, D. R. and Watson, M. T.: J. Polym. Sci., Part C, *30*, 349 (1970).
85. Onogi, S. and Asuda, T.: "Rheology, Vol. 1", Astarita, G., Marucci, G. and Nicolais, L., eds., Plenum Press, N.Y., 1980, pp. 127–187.
86. Maret, G., Blumstein, A. and Vilasagar, S.: Polymer Preprints, *22*(1), 246 (1981).
87. Jin, J.-I., Lenz, R. W. and Antoun, S.: J. Korean Chem. Soc., *26*(3), 188 (1982).
88. DeVries, A.: Mol. Cryst. Liq. Cryst., *10*, 31 (1970).
89. Azaroff, L. V.: Mol. Cryst. Liq. Cryst., *60*, 73 (1980).
90. DeVries, A.: Mol. Cryst. Liq. Cryst., *10*, 219 (1970).
91. Chandrasekhar, S.: "Liquid Crystals", Cambridge University Press, N.Y., 1977.
92. Petraccone, V., Roviello, A., Sirigu, A., Tuzi, A., Martuscelli, E. and Pracella, M.: Europ. Polym. J., *16*, 261 (1980).
93. Krigbaum, W. R., Lader, H. J. and Ciferri, A.: Macromolecules, *13*, 554 (1980).
94. Williams, R.: J. Chem. Phys., *39*, 384 (1963).

95. Heilmeir, G. H., Zanoni, L. and Barten, L.: Appl. Phys. Lett., *13*, 46 (1968).
96. Heilmeir, G. H., Zanoni, L. and Barten, L.: Proc. I.E.E.E., *56*, 1162 (1968).
97. Reference 3, Chapt. 5, pp. 110–121, Vol. 2.
98. Volino, F., Martins, A. F., Blumstein, R. B. and Blumstein, A., J. Phys. Letters (Paris), *42*, L305 (1981).
99. Reference 3, Chapt. 7, pp. 144–191, Vol. 2.
100. Vistin, L. K.: Sov. Phys. — Crystallogr. (Engl. Trans.), *15*, 514, 908 (1970).
101. Ober, C. K.: Ph.D. Thesis, University of Massachusetts, 1982.
102. Cser, F., Nytrai, K., Hardy, G., Merczel, J. and Vanga, J.: J. Polym. Sci.: Polym. Symp., *69*, 91 (1981).

M. Gordon/H.-J. Cantow (Editors)
Received March 4, 1983

Author Index Volumes 1–59

Allegra, G. and *Bassi, I. W.:* Isomorphism in Synthetic Macromolecular Systems. Vol. 6, pp. 549–574.

Andrews, E. H.: Molecular Fracture in Polymers. Vol. 27, pp. 1–66.

Anufrieva, E. V. and *Gotlib, Yu. Ya.:* Investigation of Polymers in Solution by Polarized Luminescence. Vol. 40, pp. 1–68.

Argon, A. S., Cohen, R. E., Gebizlioglu, O. S. and *Schwier, C.:* Crazing in Block Copolymers and Blends. Vol. 52/53, pp. 275–334

Arridge, R. C. and *Barham, P. J.:* Polymer Elasticity. Discrete and Continuum Models. Vol. 46, pp. 67–117.

Ayrey, G.: The Use of Isotopes in Polymer Analysis. Vol. 6, pp. 128–148.

Baldwin, R. L.: Sedimentation of High Polymers. Vol. 1, pp. 451–511.

Basedow, A. M. and *Ebert, K.:* Ultrasonic Degradation of Polymers in Solution. Vol. 22, pp. 83–148.

Batz, H.-G.: Polymeric Drugs. Vol. 23, pp. 25–53.

Bekturov, E. A. and *Bimendina, L. A.:* Interpolymer Complexes. Vol. 41, pp. 99–147.

Bergsma, F. and *Kruissink, Ch. A.:* Ion-Exchange Membranes. Vol. 2, pp. 307–362.

Berlin, Al. Al., Volfson, S. A., and *Enikolopian, N. S.:* Kinetics of Polymerization Processes. Vol. 38, pp. 89–140.

Berry, G. C. and *Fox, T. G.:* The Viscosity of Polymers and Their Concentrated Solutions. Vol. 5, pp. 261–357.

Bevington, J. C.: Isotopic Methods in Polymer Chemistry. Vol. 2, pp. 1–17.

Bhuiyan, A. L.: Some Problems Encountered with Degradation Mechanisms of Addition Polymers. Vol. 47, pp. 1–65.

Bird, R. B., Warner, Jr., H. R., and *Evans, D. C.:* Kinetik Theory and Rheology of Dumbbell Suspensions with Brownian Motion. Vol. 8, pp. 1–90.

Biswas, M. and *Maity, C.:* Molecular Sieves as Polymerization Catalysts. Vol. 31, pp. 47–88.

Block, H.: The Nature and Application of Electrical Phenomena in Polymers. Vol. 33, pp. 93–167.

Böhm, L. L., Chmelir̆, M., Löhr, G., Schmitt, B. J. and *Schulz, G. V.:* Zustände und Reaktionen des Carbanions bei der anionischen Polymerisation des Styrols. Vol. 9, pp. 1–45.

Bovey, F. A. and *Tiers, G. V. D.:* The High Resolution Nuclear Magnetic Resonance Spectroscopy of Polymers. Vol. 3, pp. 139–195.

Braun, J.-M. and *Guillet, J. E.:* Study of Polymers by Inverse Gas Chromatography. Vol. 21, pp. 107–145.

Breitenbach, J. W., Olaj, O. F. und *Sommer, F.:* Polymerisationsanregung durch Elektrolyse. Vol. 9, pp. 47–227.

Bresler, S. E. and *Kazbekov, E. N.:* Macroradical Reactivity Studied by Electron Spin Resonance. Vol. 3, pp. 688–711.

Bucknall, C. B.: Fracture and Failure of Multiphase Polymers and Polymer Composites. Vol. 27, pp. 121–148.

Burchard, W.: Static and Dynamic Light Scattering from Branched Polymers and Biopolymers. Vol. 48, pp. 1–124.

Bywater, S.: Polymerization Initiated by Lithium and Its Compounds. Vol. 4, pp. 66–110.

Bywater, S.: Preparation and Properties of Star-branched Polymers. Vol. 30, pp. 89–116.

Candau, S., Bastide, J. and *Delsanti, M.:* Structural. Elastic and Dynamic Properties of Swollen Polymer Networks. Vol. 44, pp. 27–72.

Carrick, W. L.: The Mechanism of Olefin Polymerization by Ziegler-Natta Catalysts. Vol. 12, pp. 65–86.

Casale, A. and *Porter, R. S.:* Mechanical Synthesis of Block and Graft Copolymers. Vol. 17, pp. 1–71.

Cerf, R.: La dynamique des solutions de macromolecules dans un champ de vitesses. Vol. 1, pp. 382–450.

Cesca, S., Priola, A. and *Bruzzone, M.:* Synthesis and Modification of Polymers Containing a System of Conjugated Double Bonds. Vol. 32, pp. 1–67.

Cicchetti, O.: Mechanisms of Oxidative Photodegradation and of UV Stabilization of Polyolefins. Vol. 7, pp. 70–112.

Clark, D. T.: ESCA Applied to Polymers. Vol. 24, pp. 125–188.

Coleman, Jr., L. E. and *Meinhardt, N. A.:* Polymerization Reactions of Vinyl Ketones. Vol. 1, pp. 159–179.

Comper, W. D. and *Preston, B. N.:* Rapid Polymer Transport in Concentrated Solutions. Vol. 55, pp. 105–152.

Crescenzi, V.: Some Recent Studies of Polyelectrolyte Solutions. Vol. 5, pp. 358–386.

Davydov, B. E. and *Krentsel, B. A.:* Progress in the Chemistry of Polyconjugated Systems. Vol. 25, pp. 1–46.

Dettenmaier, M.: Intrinsic Crazes in Polycarbonate Phenomenology and Molecular Interpretation of a New Phenomenon. Vol. 52/53, pp. 57–104

Döll, W.: Optical Interference Measurements and Fracture Mechanics Analysis of Crack Tip Craze Zones. Vol. 52/53, pp. 105–168

Dole, M.: Calorimetric Studies of States and Transitions in Solid High Polymers. Vol. 2, pp. 221–274.

Dreyfuss, P. and *Dreyfuss, M. P.:* Polytetrahydrofuran. Vol. 4, pp. 528–590.

Drobnik, J. and *Rypáček, F.:* Soluble Synthetic Polymers in Biological Systems. Vol. 57, pp. 1–50.

Dušek, K. and *Prins, W.:* Structure and Elasticity of Non-Crystalline Polymer Networks. Vol. 6, pp. 1–102.

Duncan, R. and *Kopeček, J.:* Soluble Synthetic Polymers as Potential Drug Carriers. Vol. 57, pp. 51–101.

Eastham, A. M.: Some Aspects of the Polymerization of Cyclic Ethers. Vol. 2, pp. 18–50.

Ehrlich, P. and *Mortimer, G. A.:* Fundamentals of the Free-Radical Polymerization of Ethylene. Vol. 7, pp. 386–448.

Eisenberg, A.: Ionic Forces in Polymers. Vol. 5, pp. 59–112.

Elias, H.-G., Bareiss, R. und *Watterson, J. G.:* Mittelwerte des Molekulargewichts und anderer Eigenschaften. Vol. 11, pp. 111–204.

Elyashevich, G. K.: Thermodynamics and Kinetics of Orientational Crystallization of Flexible-Chain Polymers. Vol. 43, pp. 207–246.

Ferruti, P. and *Barbucci, R.:* Linear Amino Polymers: Synthesis, Protonation and Complex Formation. Vol. 58, pp. 55–92.

Fischer, H.: Freie Radikale während der Polymerisation, nachgewiesen und identifiziert durch Elektronenspinresonanz. Vol. 5, pp. 463–530.

Flory, P. J.: Molecular Theory of Liquid Crystals. Vol. 59, pp. 1–36.

Ford, W. T. and *Tomoi, M.:* Polymer-Supported Phase Transfer Catalysts Reaction Mechanisms. Vol. 55, pp. 49–104.

Fradet, A. and *Maréchal, E.:* Kinetics and Mechanisms of Polyesterifications. I. Reactions of Diols with Diacids. Vol. 43, pp. 51–144.

Friedrich, K.: Crazes and Shear Bands in Semi-Crystalline Thermoplastics. Vol. 52/53, pp. 225–274

Fujita, H.: Diffusion in Polymer-Diluent Systems. Vol. 3, pp. 1–47.

Funke, W.: Über die Strukturaufklärung vernetzter Makromoleküle, insbesondere vernetzter Polyesterharze, mit chemischen Methoden. Vol. 4, pp. 157–235.

Gal'braikh, L. S. and *Rogovin, Z. A.:* Chemical Transformations of Cellulose. Vol. 14, pp. 87–130.

Gallot, B. R. M.: Preparation and Study of Block Copolymers with Ordered Structures, Vol. 29, pp. 85–156.

Gandini, A.: The Behaviour of Furan Derivatives in Polymerization Reactions. Vol. 25, pp. 47–96.

Gandini, A. and *Cheradame, H.:* Cationic Polymerization. Initiation with Alkenyl Monomers. Vol. 34/35, pp. 1–289.

Geckeler, K., Pillai, V. N. R., and *Mutter, M.:* Applications of Soluble Polymeric Supports. Vol. 39, pp. 65–94.

Gerrens, H.: Kinetik der Emulsionspolymerisation. Vol. 1, pp. 234–328.

Ghiggino, K. P., Roberts, A. J. and *Phillips, D.:* Time-Resolved Fluorescence Techniques in Polymer and Biopolymer Studies. Vol. 40, pp. 69–167.

Goethals, E. J.: The Formation of Cyclic Oligomers in the Cationic Polymerization of Heterocycles. Vol. 23, pp. 103–130.

Graessley, W. W.: The Etanglement Concept in Polymer Rheology. Vol. 16, pp. 1–179.

Graessley, W. W.: Entagled Linear, Branched and Network Polymer Systems. Molecular Theories. Vol. 47, pp. 67–117.

Hagihara, N., Sonogashira, K. and *Takahashi, S.:* Linear Polymers Containing Transition Metals in the Main Chain. Vol. 41, pp. 149–179.

Hasegawa, M.: Four-Center Photopolymerization in the Crystalline State. Vol. 42, pp. 1–49.

Hay, A. S.: Aromatic Polyethers. Vol. 4, pp. 496–527.

Hayakawa, R. and *Wada, Y.:* Piezoelectricity and Related Properties of Polymer Films. Vol. 11, pp. 1–55.

Heidemann, E. and *Roth, W.:* Synthesis and Investigation of Collagen Model Peptides. Vol. 43, pp. 145–205.

Heitz, W.: Polymeric Reagents. Polymer Design, Scope, and Limitations. Vol. 23, pp. 1–23.

Helfferich, F.: Ionenaustausch. Vol. 1, pp. 329–381.

Hendra, P. J.: Laser-Raman Spectra of Polymers. Vol. 6, pp. 151–169.

Henrici-Olivé, G. und *Olivé, S.:* Kettenübertragung bei der radikalischen Polymerisation. Vol. 2, pp. 496–577.

Henrici-Olivé, G. und *Olivé, S.:* Koordinative Polymerisation an löslichen Übergangsmetall-Katalysatoren. Vol. 6, pp. 421–472.

Henrici-Olivé, G. and *Olivé, S.:* Oligomerization of Ethylene with Soluble Transition-Metal Catalysts. Vol. 15, pp. 1–30.

Henrici-Olivé, G. and *Olivé, S.:* Molecular Interactions and Macroscopic Properties of Polyacrylonitrile and Model Substances. Vol. 32, pp. 123–152.

Henrici-Olivé, G. and *Olivé, S.:* The Chemistry of Carbon Fiber Formation from Polyacrylonitrile. Vol. 51, pp. 1–60.

Hermans, Jr., J., Lohr, D. and *Ferro, D.:* Treatment of the Folding and Unfolding of Protein Molecules in Solution According to a Lattic Model. Vol. 9, pp. 229–283.

Hoffman, A. S.: Ionizing Radiation and Gas Plasma (or Glow) Discharge Treatments for Preparation of Novel Polymeric Biomaterials. Vol. 57, pp. 141–157.

Holzmüller, W.: Molecular Mobility, Deformation and Relaxation Processes in Polymers. Vol. 26, pp. 1–62.

Hutchison, J. and *Ledwith, A.:* Photoinitiation of Vinyl Polymerization by Aromatic Carbonyl Compounds. Vol. 14, pp. 49–86.

Iizuka, E.: Properties of Liquid Crystals of Polypeptides: with Stress on the Electromagnetic Orientation. Vol. 20, pp. 79–107.

Ikada, Y.: Characterization of Graft Copolymers. Vol. 29, pp. 47–84.

Ikada, Y.: Blood-Compatible Polymers. Vol. 57, pp. 103–140.

Imanishi, Y.: Synthese, Conformation, and Reactions of Cyclic Peptides. Vol. 20, pp. 1–77.

Inagaki, H.: Polymer Separation and Characterization by Thin-Layer Chromatography. Vol. 24, pp. 189–237.

Inoue, S.: Asymmetric Reactions of Synthetic Polypeptides. Vol. 21, pp. 77–106.

Ise, N.: Polymerizations under an Electric Field. Vol. 6, pp. 347–376.

Ise, N.: The Mean Activity Coefficient of Polyelectrolytes in Aqueous Solutions and Its Related Properties. Vol. 7, pp. 536–593.

Isihara, A.: Intramolecular Statistics of a Flexible Chain Molecule. Vol. 7, pp. 449–476.

Isihara, A.: Irreversible Processes in Solutions of Chain Polymers. Vol. 5, pp. 531–567.

Isihara, A. and *Guth, E.:* Theory of Dilute Macromolecular Solutions. Vol. 5, pp. 233–260.

Iwatsuki, S.: Polymerization of Quinodimethane Compounds. Vol. 58, pp. 93–120.

Janeschitz-Kriegl, H.: Flow Birefrigence of Elastico-Viscous Polymer Systems. Vol. 6, pp. 170–318.

Jenkins, R. and *Porter, R. S.:* Upertubed Dimensions of Stereoregular Polymers. Vol. 36, pp. 1–20.

Jenngins, B. R.: Electro-Optic Methods for Characterizing Macromolecules in Dilute Solution. Vol. 22, pp. 61–81.

Johnston, D. S.: Macrozwitterion Polymerization. Vol. 42, pp. 51–106.

Kamachi, M.: Influence of Solvent on Free Radical Polymerization of Vinyl Compounds. Vol. 38, pp. 55–87.

Kaneko, M. and *Yamada, A.:* Solar Energy Conversion by Functional Polymers. Vol. 55, pp. 1–48.

Kawabata, S. and *Kawai, H.:* Strain Energy Density Functions of Rubber Vulcanizates from Biaxial Extension. Vol. 24, pp. 89–124.

Kennedy, J. P. and *Chou, T.:* Poly(isobutylene-co-β-Pinene): A New Sulfur Vulcanizable, Ozone Resistant Elastomer by Cationic Isomerization Copolymerization. Vol. 21, pp. 1–39.

Kennedy, J. P. and *Delvaux, J. M.:* Synthesis, Characterization and Morphology of Poly(butadiene-g-Styrene). Vol. 38, pp. 141–163.

Kennedy, J. P. and *Gillham, J. K.:* Cationic Polymerization of Olefins with Alkylaluminium Initiators. Vol. 10, pp. 1–33.

Kennedy, J. P. and *Johnston, J. E.:* The Cationic Isomerization Polymerization of 3-Methyl-1-butene and 4-Methyl-1-pentene. Vol. 19, pp. 57–95.

Kennedy, J. P. and *Langer, Jr., A. W.:* Recent Advances in Cationic Polymerization. Vol. 3, pp. 508–580.

Kennedy, J. P. and *Otsu, T.:* Polymerization with Isomerization of Monomer Preceding Propagation. Vol. 7, pp. 369–385.

Kennedy, J. P. and *Rengachary, S.:* Correlation Between Cationic Model and Polymerization Reactions of Olefins. Vol. 14, pp. 1–48.

Kennedy, J. P. and *Trivedi, P. D.:* Cationic Olefin Polymerization Using Alkyl Halide — Alkylaluminium Initiator Systems. I. Reactivity Studies. II. Molecular Weight Studies. Vol. 28, pp. 83–151.

Kennedy, J. P., Chang, V. S. C. and *Guyot, A.:* Carbocationic Synthesis and Characterization of Polyolefins with Si–H and Si–Cl Head Groups. Vol. 43, pp. 1–50.

Khoklov, A. R. and *Grosberg, A. Yu.* Statistical Theory of Polymeric Lyotropic Liquid Crystals. Vol. 41, pp. 53–97.

Kissin, Yu. V.: Structures of Copolymers of High Olefins. Vol. 15, pp. 91–155.

Kitagawa, T. and *Miyazawa, T.:* Neutron Scattering and Normal Vibrations of Polymers. Vol. 9, pp. 335–414.

Kitamaru, R. and *Horii, F.:* NMR Approach to the Phase Structure of Linear Polyethylene. Vol. 26, pp. 139–180.

Knappe, W.: Wärmeleitung in Polymeren. Vol. 7, pp. 477–535.

Koenig, J. L.: Fourier Transforms Infrared Spectroscopy of Polymers, Vol. 54, pp. 87–154.

Kolařik, J.: Secondary Relaxations in Glassy Polymers: Hydrophilic Polymethacrylates and Polyacrylates: Vol. 46, pp. 119–161.

Koningsveld, R.: Preparative and Analytical Aspects of Polymer Fractionation. Vol. 7.

Kovacs, A. J.: Transition vitreuse dans les polymers amorphes. Etude phénoménologique. Vol. 3, pp. 394–507.

Krässig, H. A.: Graft Co-Polymerization of Cellulose and Its Derivatives. Vol. 4, pp. 111–156.

Kramer, E. J.: Microscopic and Molecular Fundamentals of Crazing. Vol. 52/53, pp. 1–56.

Kraus, G.: Reinforcement of Elastomers by Carbon Black. Vol. 8, pp. 155–237.

Kreutz, W. and *Welte, W.:* A General Theory for the Evaluation of X-Ray Diagrams of Biomembranes and Other Lamellar Systems. Vol. 30, pp. 161–225.

Krimm, S.: Infrared Spectra of High Polymers. Vol. 2, pp. 51–72.

Kuhn, W., Ramel, A., Walters, D. H., Ebner, G. and *Kuhn, H. J.:* The Production of Mechanical Energy from Different Forms of Chemical Energy with Homogeneous and Cross-Striated High Polymer Systems. Vol. 1, pp. 540–592.

Kunitake, T. and *Okahata, Y.:* Catalytic Hydrolysis by Synthetic Polymers. Vol. 20, pp. 159–221.

Kurata, M. and *Stockmayer, W. H.:* Intrinsic Viscosities and Unperturbed Dimensions of Long Chain Molecules. Vol. 3, pp. 196–312.

Ledwith, A. and *Sherrington, D. C.:* Stable Organic Cation Salts: Ion Pair Equilibria and Use in Cationic Polymerization. Vol. 19, pp. 1–56.

Lee, C.-D. S. and *Daly, W. H.:* Mercaptan-Containing Polymers. Vol. 15, pp. 61–90.

Lipatov, Y. S.: Relaxation and Viscoelastic Properties of Heterogeneous Polymeric Compositions. Vol. 22, pp. 1–59.

Lipatov, Y. S.: The Iso-Free-Volume State and Glass Transitions in Amorphous Polymers: New Development of the Theory. Vol. 26, pp. 63–104.

Lustoň, J. and *Vašš, F.:* Anionic Copolymerization of Cyclic Ethers with Cyclic Anhydrides. Vol. 56, pp. 91–133.

Mano, E. B. and *Coutinho, F. M. B.:* Grafting on Polyamides. Vol. 19, pp. 97–116.

Mark, J. E.: The Use of Model Polymer Networks to Elucidate Molecular Aspects of Rubberlike Elasticity. Vol. 44, pp. 1–26.

Meerwall v., E., D.: Self-Diffusion in Polymer Systems, Measured with Field-Gradient Spin Echo NMR Methods, Vol. 54, pp. 1–29.

Mengoli, G.: Feasibility of Polymer Film Coating Through Electroinitiated Polymerization in Aqueous Medium. Vol. 33, pp. 1–31.

Meyerhoff, G.: Die viscosimetrische Molekulargewichtsbestimmung von Polymeren. Vol. 3, pp. 59–105.

Millich, F.: Rigid Rods and the Characterization of Polyisocyanides. Vol. 19, pp. 117–141.

Morawetz, H.: Specific Ion Binding by Polyelectrolytes. Vol. 1, pp. 1–34.

Morin, B. P., Breusova, I. P. and *Rogovin, Z. A.:* Structural and Chemical Modifications of Cellulose by Graft Copolymerization. Vol. 42, pp. 139–166.

Mulvaney, J. E., Oversberger, C. C. and *Schiller, A. M.:* Anionic Polymerization. Vol. 3, pp. 106–138.

Neuse, E.: Aromatic Polybenzimidazoles. Syntheses, Properties, and Applications. Vol. 47, pp. 1–42.

Ober, Ch. K., Jin, J.-I. and *Lenz, R. W.:* Liquid Crystal Polymers with Flexible Spacers in the Main Chain. Vol. 59, pp. 103–146.

Okubo, T. and *Ise, N.:* Synthetic Polyelectrolytes as Models of Nucleic Acids and Esterases. Vol. 25, pp. 135–181.

Osaki, K.: Viscoelastic Properties of Dilute Polymer Solutions. Vol. 12, pp. 1–64.

Oster, G. and *Nishijima, Y.:* Fluorescence Methods in Polymer Science. Vol. 3, pp. 313–331.

Overberger, C. G. and *Moore, J. A.:* Ladder Polymers. Vol. 7, pp. 113–150.

Papkov, S. P.: Liquid Crystalline Order in Solutions of Rigid-Chain Polymers. Vol. 59, pp. 75–102.

Patat, F., Killmann, E. und *Schiebener, C.:* Die Absorption von Makromolekülen aus Lösung. Vol. 3, pp. 332–393.

Patterson, G. D.: Photon Correlation Spectroscopy of Bulk Polymers. Vol. 48, pp. 125–159.

Penczek, S., Kubisa, P. and *Matyjaszewski, K.:* Cationic Ring-Opening Polymerization of Heterocyclic Monomers. Vol. 37, pp. 1–149.

Peticolas, W. L.: Inelastic Laser Light Scattering from Biological and Synthetic Polymers. Vol. 9, pp. 285–333.

Pino, P.: Optically Active Addition Polymers. Vol. 4, pp. 393–456.

Pitha, J.: Physiological Activities of Synthetic Analogs of Polynucleotides. Vol. 50, pp. 1–16.

Plate, N. A. and *Noah, O. V.:* A Theoretical Consideration of the Kinetics and Statistics of Reactions of Functional Groups of Macromolecules. Vol. 31, pp. 133–173.

Plesch, P. H.: The Propagation Rate-Constants in Cationic Polymerisations. Vol. 8, pp. 137–154.

Porod, G.: Anwendung und Ergebnisse der Röntgenkleinwinkelstreuung in festen Hochpolymeren. Vol. 2, pp. 363–400.

Pospíšil, J.: Transformations of Phenolic Antioxidants and the Role of Their Products in the Long-Term Properties of Polyolefins. Vol. 36, pp. 69–133.

Postelnek, W., Coleman, L. E., and *Lovelace, A. M.:* Fluorine-Containing Polymers. I. Fluorinated Vinyl Polymers with Functional Groups, Condensation Polymers, and Styrene Polymers. Vol. 1, pp. 75–113.

Rempp, P. F. and *Franta, E.:* Macromonomers: Synthesis, Characterization and Applications. Vol. 58, pp. 1–54.

Rempp, P., Herz, J., and *Borchard, W.:* Model Networks. Vol. 26, pp. 107–137.

Rigbi, Z.: Reinforcement of Rubber by Carbon Black. Vol. 36, pp. 21–68.

Rogovin, Z. A. and *Gabrielyan, G. A.:* Chemical Modifications of Fibre Forming Polymers and Copolymers of Acrylonitrile. Vol. 25, pp. 97–134.

Roha, M.: Ionic Factors in Steric Control. Vol. 4, pp. 353–392.

Roha, M.: The Chemistry of Coordinate Polymerization of Dienes. Vol. 1, pp. 512–539.

Safford, G. J. and *Naumann, A. W.:* Low Frequency Motions in Polymers as Measured by Neutron Inelastic Scattering. Vol. 5, pp. 1–27.

Sauer, J. A. and *Chen, C. C.:* Crazing and Fatigue Behavior in One and Two Phase Glassy Polymers. Vol. 52/53, pp. 169–224

Schuerch, C.: The Chemical Synthesis and Properties of Polysaccharides of Biomedical Interest. Vol. 10, pp. 173–194.

Schulz, R. C. und *Kaiser, E.:* Synthese und Eigenschaften von optisch aktiven Polymeren. Vol. 4, pp. 236–315.

Seanor, D. A.: Charge Transfer in Polymers. Vol. 4, pp. 317–352.

Semerak, S. N. and *Frank, C. W.:* Photophysics of Excimer Formation in Aryl Vinyl Polymers, Vol. 54, pp. 31–85.

Seidl, J., Malinský, J., Dušek, K. und *Heitz, W.:* Makroporöse Styrol-Divinylbenzol-Copolymere und ihre Verwendung in der Chromatographie und zur Darstellung von Ionenaustauschern. Vol. 5, pp. 113–213.

Semjonow, V.: Schmelzviskositäten hochpolymerer Stoffe. Vol. 5, pp. 387–450.

Semlyen, J. A.: Ring-Chain Equilibria and the Conformations of Polymer Chains. Vol. 21, pp. 41–75.

Sharkey, W. H.: Polymerizations Through the Carbon-Sulphur Double Bond. Vol. 17, pp. 73–103.

Shimidzu, T.: Cooperative Actions in the Nucleophile-Containing Polymers. Vol. 23, pp. 55–102.

Shutov, F. A.: Foamed Polymers Based on Reactive Oligomers, Vol. 39, pp. 1–64.

Shutov, F. A.: Foamed Polymers. Cellular Structure and Properties. Vol. 51, pp. 155–218.

Silvestri, G., Gambino, S., and *Filardo, G.:* Electrochemical Production of Initiators for Polymerization Processes. Vol. 38, pp. 27–54.

Slichter, W. P.: The Study of High Polymers by Nuclear Magnetic Resonance. Vol. 1, pp. 35–74.

Small, P. A.: Long-Chain Branching in Polymers. Vol. 18.

Smets, G.: Block and Graft Copolymers. Vol. 2, pp. 173–220.

Smets, G.: Photochromic Phenomena in the Solid Phase. Vol. 50, pp. 17–44.

Sohma, J. and *Sakaguchi, M.:* ESR Studies on Polymer Radicals Produced by Mechanical Destruction and Their Reactivity. Vol. 20, pp. 109–158.

Sotobayashi, H. und *Springer, J.:* Oligomere in verdünnten Lösungen. Vol. 6, pp. 473–548.

Sperati, C. A. and *Starkweather, Jr., H. W.:* Fluorine-Containing Polymers. II. Polytetrafluoroethylene. Vol. 2, pp. 465–495.

Sprung, M. M.: Recent Progress in Silicone Chemistry. I. Hydrolysis of Reactive Silane Intermediates, Vol. 2, pp. 442–464.

Stahl, E. and *Brüderle, V.:* Polymer Analysis by Thermofractography. Vol. 30, pp. 1–88.

Stannett, V. T., Koros, W. J., Paul, D. R., Lonsdale, H. K., and *Baker, R. W.:* Recent Advances in Membrane Science and Technology. Vol. 32, pp. 69–121.

Staverman, A. J.: Properties of Phantom Networks and Real Networks. Vol. 44, pp. 73–102.

Stauffer, D., Coniglio, A. and *Adam, M.:* Gelation and Critical Phenomena. Vol. 44, pp. 103–158.

Stille, J. K.: Diels-Alder Polymerization. Vol. 3, pp. 48–58.

Stolka, M. and *Pai, D.:* Polymers with Photoconductive Properties. Vol. 29, pp. 1–45.

Subramanian, R. V.: Electroinitiated Polymerization on Electrodes. Vol. 33, pp. 35–58.

Sumitomo, H. and *Okada, M.:* Ring-Opening Polymerization of Bicyclic Acetals, Oxalactone, and Oxalactam. Vol. 28, pp. 47–82.

Szegő, L.: Modified Polyethylene Terephthalate Fibers. Vol. 31, pp. 89–131.

Szwarc, M.: Termination of Anionic Polymerization. Vol. 2, pp. 275–306.

Szwarc, M.: The Kinetics and Mechanism of N-carboxy-α-amino-acid Anhydride (NCA) Polymerization to Poly-amino Acids. Vol. 4, pp. 1–65.

Szwarc, M.: Thermodynamics of Polymerization with Special Emphasis on Living Polymers. Vol. 4, pp. 457–495.

Szwarc, M.: Living Polymers and Mechanisms of Anionic Polymerization. Vol. 49, pp. 1–175.

Takahashi, A. and *Kawaguchi, M.:* The Structure of Macromolecules Adsorbed on Interfaces. Vol. 46, pp. 1–65.

Takemoto, K. and *Inaki, Y.:* Synthetic Nucleic Acid Analogs. Preparation and Interactions. Vol. 41, pp. 1–51.

Tani, H.: Stereospecific Polymerization of Aldehydes and Epoxides. Vol. 11, pp. 57–110.

Tate, B. E.: Polymerization of Itaconic Acid and Derivatives. Vol. 5, pp. 214–232.

Tazuke, S.: Photosensitized Charge Transfer Polymerization. Vol. 6, pp. 321–346.

Teramoto, A. and *Fujita, H.:* Conformation-dependent Properties of Synthetic Polypeptides in the Helix-Coil Transition Region. Vol. 18, pp. 65–149.

Thomas, W. M.: Mechanismus of Acrylonitrile Polymerization. Vol. 2, pp. 401–441.

Tobolsky, A. V. and *DuPré, D. B.:* Macromolecular Relaxation in the Damped Torsional Oscillator and Statistical Segment Models. Vol. 6, pp. 103–127.

Tosi, C. and *Ciampelli, F.:* Applications of Infrared Spectroscopy to Ethylene-Propylene Copolymers. Vol. 12, pp. 87–130.

Tosi, C.: Sequence Distribution in Copolymers: Numerical Tables. Vol. 5, pp. 451–462.

Tsuchida, E. and *Nishide, H.:* Polymer-Metal Complexes and Their Catalytic Activity. Vol. 24, pp. 1–87.

Tsuji, K.: ESR Study of Photodegradation of Polymers. Vol. 12, pp. 131–190.

Tsvetkov, V. and *Andreeva, L.:* Flow and Electric Birefringence in Rigid-Chain Polymer Solutions. Vol. 39, pp. 95–207.

Tuzar, Z., Kratochvil, P., and *Bohdanecký, M.:* Dilute Solution Properties of Aliphatic Polyamides. Vol. 30, pp. 117–159.

Uematsu, I. and *Uematsu, Y.:* Polypeptide Liquid Crystals. Vol. 59, pp. 37–74.

Valvassori, A. and *Sartori, G.:* Present Status of the Multicomponent Copolymerization Theory. Vol. 5, pp. 28–58.

Voorn, M. J.: Phase Seperation in Polymer Solutions. Vol. 1, pp. 192–233.

Werber, F. X.: Polymerization of Olefins on Supported Catalysts. Vol. 1, pp. 180–191.

Wichterle, O., Šebenda, J., and *Králiček, J.:* The Anionic Polymerization of Caprolactam. Vol. 2, pp. 578–595.

Wilkes, G. L.: The Measurement of Molecular Orientation in Polymeric Solids. Vol. 8, pp. 91–136.

Williams, G.: Molecular Aspects of Multiple Dielectric Relaxation Processes in Solid Polymers. Vol. 33, pp. 59–92.

Williams, J. G.: Applications of Linear Fracture Mechanics. Vol. 27, pp. 67–120.

Wöhrle, D.: Polymere aus Nitrilen. Vol. 10, pp. 35–107.

Wöhrle, D.: Polymer Square Planar Metal Chelates for Science and Industry. Synthesis, Properties and Applications. Vol. 50, pp. 45–134.

Wolf, B. A.: Zur Thermodynamik der enthalpisch und der entropisch bedingten Entmischung von Polymerlösungen. Vol. 10, pp. 109–171.

Woodward, A. E. and *Sauer, J. A.:* The Dynamic Mechanical Properties of High Polymers at Low Temperatures. Vol. 1, pp. 114–158.

Wunderlich, B. and *Baur, H.:* Heat Capacities of Linear High Polymers. Vol. 7, pp. 151–368.

Wunderlich, B.: Crystallization During Polymerization. Vol. 5, pp. 568–619.

Wrasidlo, W.: Thermal Analysis of Polymers. Vol. 13, pp. 1–99.

Yamashita, Y.: Random and Black Copolymers by Ring-Opening Polymerization. Vol. 28, pp. 1–46.

Yamazaki, N.: Electrolytically Initiated Polymerization. Vol. 6, pp. 377–400.

Yamazaki, N. and *Higashi, F.:* New Condensation Polymerizations by Means of Phosphorus Compounds. Vol. 38, pp. 1–25.

Yokoyama, Y. and *Hall, H. K.:* Ring-Opening Polymerization of Atom-Bridged and Bond-Bridged Bicyclic Ethers, Acetals and Orthoesters. Vol. 42, pp. 107–138.

Yoshida, H. and *Hayashi, K.:* Initiation Process of Radiation-induced Ionic Polymerization as Studied by Electron Spin Resonance. Vol. 6, pp. 401–420.

Young, R. N., Quirk, R. P. and *Fetters, L. J.:* Anionic Polymerizations of Non-Polar Monomers Involving Lithium. Vol. 56, pp. 1–90.

Yuki, H. and *Hatada, K.:* Stereospecific Polymerization of Alpha-Substituted Acrylic Acid Esters. Vol. 31, pp. 1–45.

Zachmann, H. G.: Das Kristallisations- und Schmelzverhalten hochpolymerer Stoffe. Vol. 3, pp. 581–687.

Zakharov, V. A., Bukatov, G. D., and *Yermakov, Y. I.:* On the Mechanism of Olifin Polymerization by Ziegler-Natta Catalysts. Vol. 51, pp. 61–100.

Zambelli, A. and *Tosi, C.:* Stereochemistry of Propylene Polymerization. Vol. 15, pp. 31–60.

Zucchini, U. and *Cecchin, G.:* Control of Molecular-Weight Distribution in Polyolefins Synthesized with Ziegler-Natta Catalytic Systems. Vol. 51, pp. 101–154.

Subject Index

Alkyl substituents 118
Anisotropic phase 8
— properties of the system 89
Anisotropy of the polarizability 2, 28
— of the polarizability,
    orientation-dependent 28
Approximate treatment, partition
    function 34
Artifically created (thermodynamically
    unstable) oriented state 82
Asymmetrie of macromolecules 77
Axial ratio 4, 11, 19, 29, 118
Azo link 112
— polymers 113
Azoxy link 112
— polymers 113

Benzene-ring stack 45
Bicyclohexylene groups 114
Binaphthol 116
Biphasic equilibria in lyotropic systems 30
— equilibrium 9, 10
— gap 13, 30
— regions 41
Biphenyl 114, 116
p, p′-Biphenylene 111
Bragg's law 45
Bulkiness 112

Capillary rheometer 140
Carbonate linkage 122
Cellulose 18
— acetate 20
— ethers 21
— derivatives 18, 20
—, hydroxypropyl 20
Characteristic temperature 28
Chemical potentials 9, 29, 33, 35
Chiral 129
Cholesteric 128, 129, 130, 132, 133, 135
— liquid crystals 38
— liquids 2
— phase separation, in cellulose derivates 20

Circular dichroism 48
— polarized dichroic ratio D 48
Coil-helix (c → h) transition 25
Combinatory partition function 5
Compensation phenomena 59
"Complex" phase 43
Compliance 142
Conformational changes 24
— effects 124
Conjugated structure 113
Conjugation 117
"Continuous-phase" morphology 23
Cooperative conformational transition 25
Copolyesters 129, 132
Copolymers 23, 127, 129, 130, 131
Crystallinity 138
Critical volume fraction, metastable order 8
Cybotactic 135

Decamethylene spacer 110, 113, 115
Degree of asymmetry of macromolecules 77
Dependence of the viscosity on the shear
    stress 89
Depolarized light intensity 134
De Vries theory 48
Die swell 141
Diformyl 113
Dipolar effects 114
Dipole moment 117
Dipole-quadrupol interactions 56
Direct linkage 117
Disorder 7
Dispersion forces 28
DSC 138
DTA 138
Dyad 112, 114

Effect of a magnetic field 90
— of electric field 91
— of external mechanical fields 93
Electrical field 143
Enthalpy of clearing 126
— of the transitions 138
Entropies of clearing 122

Entropy of clearing 123, 139
Equilibrium between two anisotropic
   phases 26
Ester dyads 109
— group 112, 114
— interchange polymerization 109
— link 109, 111
— unit 113
Ether 111
Ethylene terephthalate (ET) 23
Even-odd behavior 123
— effect 139
— relationship 121, 122
Exact treatment 31

Fan-shaped 133
Flexible spacer 104, 105, 106, 116, 120
— unit 106
Flory theory 39
— —, point A 38
— —, point B 38
Focal-conic 133
— textures 125
Formation of crystallosolvates 91
Fractionation of species 15

Gauche conformations 124
Geometrical (topological) analysis of phase
   equilibria 88
Geometries 111
Glass transitions 127

α-Helical polypeptides, in liquid
   crystallinity 11
Homeotropic 133, 134
Homologous series 120, 123
Homopolymer 129
Hot-stage 132
p-Hydroxybenzoate (PHB) 23

Imino links 113
Incipient separation of the nematic phase 10
Increased spacer length 121
Intermolecular attractive forces 25
Iridescent 129
— colors 47
Isotropic 111, 133
— forces 25
— phase 15, 124
— -nematic transition 24
Isomorphous 125

Keating's theory 52
Kinetics of phase transitions 95
Kuhn chain 19

Lateral substituents 117
Lattice methods 4
— theory 3
— — for hard rods 31
— — of liquid crystallinity 3
— — of orientational order 3
Length of the mesogenic group 114
Length-to-diameter ratio 116
Length-to-width ratio 114
Leslie-Erickson theory 142
Linking groups 105, 106
Liquid crystal induced CD, LCICD 49
Liquid crystals 1
Liquid crystallinity 2, 12
— — semiflexible chains 18
Lyotropic liquid crystalline systems 81
— — crystals 76
— systems 4, 11

Magnetic field 143
Maier-Saupe theory 2, 124
Melting temperature 114, 127
Melt viscosity 141
Mesogenic group 105, 106, 115
Mesomorphic behavior of random
   copolymers 23
Mesophase stability 116, 122, 127
3-Methyl-adipic 129
3-Methyl adipic acid 132
Micro-domains of the nematic phase 23
Microscopic 132
Miscibility 134
Molecular polarizability 124
— theory 1
— — of the liquid crystalline state 83
— — of liquid crystals 1
— weight 110
Monotropic 112, 126, 131, 137
Multicomponent systems theory 14

Nature of the solvent on phase
   equilibrium 93
Nematic 111, 112, 113, 114, 127, 128, 133,
   135
— copolymers 24
— liquid crystalline state 2
— liquid crystals 2
— liquids 2
— phase 15
— polyesters 121
— state 126
— temperature T_N 54

Oily streaks 133
— streak textures 130
Onogi-Asada model 142
Optically active 128

Optical rotation 48
Order 7
— -disorder transition 7
— parameter 124, 136, 143
Orientation 136
— -dependent energy 28
— — forces 2, 26
— — interactions 27, 28, 34
— distribution 31
Orientational partition function 6
Oscillation of transition temperatures 22

Para-aromatic polyamides 78
Parallel disclinations 133
Para positions 109, 117
Partition function 6
— — for hard rodlike particles 4
Phase diagram 134
— equilibrium 81
— separation 38
— transition conformational transition 25
— —, coupled 25
p-Phenylene ring 111
Planarity and rigidity of the mesogenic
    unit 114
Polarity 116
Polarizability 117, 143
Polarized light microscopy 132
Polyaramides 15
Poly(γ-benzyl-L-glutamate) (PBLG) 78
Polycarbonates 126
Polyester 117
Poly(ethylene oxide) spacers 125
Polydisperse rods 14
Polydispersity, effect of 13
Polyisocyanates 15
Polymeric aramides 12
Polymers consisting of rigid and flexible
    units 22
Polymethylene 120
— spacers 125
Poly(N-alkyl isocyanate) (PIC) 13
Poly(p-benzamide) (PBA) 12, 78
Poly(phenylene benzothiazole) (PBT) 17
Poly(p-phenylene terephthalamide)
    (PPDT) 12, 78
Polypeptide 15, 38
Poly(propylene oxide) 128
Polysaccharides, liquid crystalline phases 21
Polysiloxane spacers 127
P-oxybenzoate and hydroquinone units 109
Pretransitional regions 40

Quenching 136

Racemic 129
Rate of appearance of a new phase 96

Reductions of the transition
    temperatures 121
Rheological behavior 89
Rheology 140
Right-handed cholesteric liquid crystal 48
Rigid-chain polymer-solvent system 81
— group 117
— -rod theory 124
Rodlike molecules 2
Rods and random coils, mixtures 16

Sample history 110, 138
Schiff base 113
— — linkage 114
Schizophyllan 22
Schlieren 133
— texture 123, 127
Selective reflection 48
Semiflexible lattice 19
Semirigid chains 18
Shear 140
— thinning 142
Smectic 111, 112, 113, 126, 133
— A 114
— B 137
— C 114, 125, 135
— E 114, 125, 136, 137
— mesophase 121
"Soft" intermolecular interactions 25
Solubility of rigid-chain polymers 80
Spontaneous elongation 95
Steric effect 119
Stir opalescence 134
Storage modulus 142
Substituents 112, 117, 128
Substitution 113
Superhelices 44
Surface treatments 133

Terephthalate 109
Terminal functional group 112
Terphenyl 116, 125
Tetraethylene oxide spacer 126
Texture 132, 133
Theoretical treatment of polydisperse
    systems 14
Thermal analysis 134, 137
Thermally-induced inversion of the
    cholesteric sense 55
Thermotropically mesomorphic polymers 69
Thermotropic liquid crystals 77
— melts 22
Therphenyl 113
Trans conformation 124
Transesterification 118
Transition of semirigid-chain polymers into
    the liquid crystalline 98

Trans-stilbene 113
Trans-vinylene 113
Triads 109, 114
Triad polymers 111
— mesogenic unit 112
True (thermodynamically stable) anisotropic
    state 82
Twisted smetic C 135
Twisting power 55

Viscosity 140

Williams domain 143
Wormlike model, chain 18

X-ray diffraction 125, 126, 135
Xylan 21